CLINICAL GENETICS

MADE RIDICULOUSLY SIMPLE

Stephen Goldberg, M.D.
Professor Emeritus
University of Miami
Miller School of Medicine
Miami, Florida

ISBN #978-1-935660-43-9

Made in the United States of America

Published by Medmaster, Inc.
P.O. Box 640028
Miami, FL 33164

Contents

Part II. When Things Go Wrong

Part III. Diagnosis and Treatment of Genetic Disorders

Introduction

Two rules in medicine:

1. Treat the cause of the problem when practicable.
2. Otherwise, treat symptomatically.

VINDICATE ME is a mnemonic for potential causes of a disease:

V: Vascular
I: Inflammatory/Infectious
N: Neoplastic
D: Degenerative
I: Intoxication, drugs
C: Congenital
A: Allergic/Autoimmune
T: Traumatic
E: Endocrine
ME: Metabolic
EM: Emotional (Psychiatric)

Where does genetics fit in here? It's there, but not shown because, digging down further, genetics can be at the root of most of the categories in VINDICATE ME: Genetic defects (i.e. those arising in the genes) can cause vascular problems, inflammatory problems, neoplastic problems, degenerative problems, drug-induced problems, allergic and autoimmune problems, endocrine problems, metabolic dysfunction, and a number of psychiatric problems. "Congenital" on the list does not necessarily imply Genetics, since congenital problems (those found at birth) need not be genetic; they may result from infection, trauma, drugs or intoxicants during pregnancy. Conversely, genetic problems need not be congenital; alterations of the genes may arise after birth, e.g. most cancers.

Genetics, then, may be at the root of many of the categories of VINDICATE ME; and if you want to attack the cause of the problem, consider genetics as the root of many problems.

But how can you treat a genetic problem? The genes seem so inaccessible.

First, there is more than just treating a disease. Patients who have a family history of the condition may want to know the likelihood that they or their children will be affected, or whether they should have children at all. Examining a family tree and arriving at a diagnosis and a prognosis for the future inheritance of a disease are important aspects of therapy, which sometimes is supportive, involving psychologic counseling, and physical and occupational therapy.

Second, today there is much that can be done medically to treat genetic diseases. When a mutation causes the buildup of a toxic chemical, it may be possible to remove the toxic substance. If the mutation causes a deficiency of an important product, it may be possible to provide the patient with the needed product (**Figure 1-1**). We have surgery for a number of congenital malformations and drugs to correct immune system dysfunctions, hormonal imbalances, clotting disorders and numerous other problems that originate in the genes. There is also the burgeoning field of gene therapy, where corrections can be made directly in the genes.

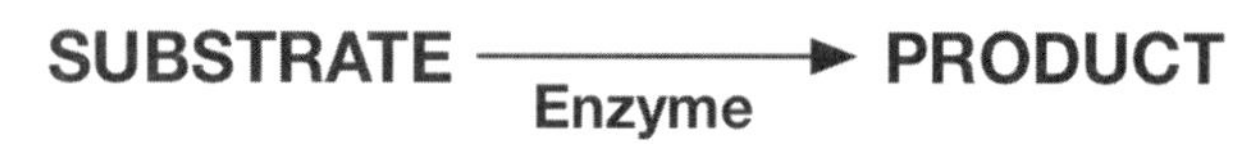

Fig. 1-1.

Figure 1-1. An **enzyme** is a protein that facilitates a chemical reaction without itself being changed by the reaction. A defective enzyme may lead to the toxic backup of a substrate that is not processed. Or it may lead to a deficiency in a needed product. Clinical methods may often be used to remove the excess substrate or provide the needed product. For instance, an enzyme defect that causes buildup of the substrate phenylalanine in *phenylketonuria,* a cause of neurological problems, may be treated with dietary restriction of phenylalanine. A defect in the production of a needed steroid product in *congenital adrenal hyperplasia* can be treated by administering steroids.

It is easy to get lost in the forest of over 6,000 genetic diseases and not see the overall picture of diagnosis and treatment. This book is not a reference text. It provides a brief overview of clinical genetics so as not to get lost in the forest through the trees. The bulk of the text focuses on general principles, with more detail in the Appendix on the several hundred individual diseases mentioned in the text and their treatment.

Part I. Classic Genetics

Mendelian Genetics

Mendelian genetics refers to several simple ideas that arose from the experiments of Gregor Mendel in the 1860's. Before Mendel's experiments, many people believed that inheritance was based on the permanent homogeneous mixing of "essences" from the parents, such as the blending of a dark skin color trait with a light skin color trait resulting in a light brown skin color, or a red flower with a white flower resulting in a pink flower. Simple observation seemed to confirm this, but Mendel believed that inheritance resulted from the combining of discrete hereditary units that remained unchanged in the offspring rather than blending together. Those same isolated units could reappear intact in subsequent generations.

Mendel concluded this when he crossed tall (T) pea plants with short (t) ones. The entire first generation was tall, with no short plants. When he crossed the first generation with itself, though, the second generation resulted in 75% tall and 25% short. This led Mendel to believe that there were two factors for height, one for tall and one for short. Tall was *dominant* over short. In order to get a short plant, the plant needed to have two short factors. A plant would be tall if it either had two tall factors (TT), or one tall factor and one small (Tt), because tall was dominant. Thus, a cross of TT x tt resulted in all talls (**Figure 1-2**).

Figure 1-2. Simple Mendelian inheritance. A recessive trait (t for short), which is not expressed in the first generation cross, reappears in 25% of the progeny in the second generation. The dominant trait (T for tall) is expressed wherever if appears.

What Mendel called *factors*, we now know as *genes*. Mendel derived three rules (using today's terminology of genes):

1. *Genes can be dominant or recessive* (**Figure 1-2**). A **genotype** refers to the genes that a person carries

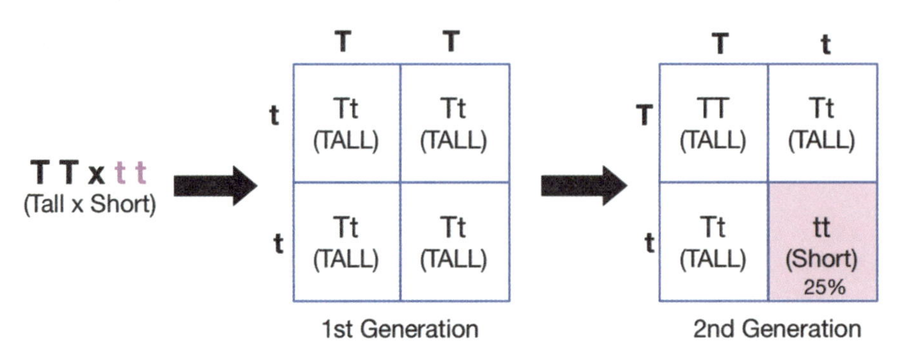

Fig. 1-2.

(also called the **genome**). A **phenotype** is how the gene expresses itself, i.e., the observable trait in the person that results from the expression of the gene. For instance, a brown-eyed phenotype results from a gene for brown eyes. A gene is **dominant** if it is expressed when the person has either one or both copies of the gene. A gene is **recessive** if it requires both copies of the gene for the gene to be expressed. For instance, the gene for brown eyes (we'll call it B) is dominant over the gene for blue eyes (we'll call it b). If a person's gene pair is BB or Bb, the person will be brown-eyed, because B is dominant over b. The b gene is recessive. The only way to have blue eyes is if both of the genes for eye color are b (bb). BB (brown-eyed) and bb (blue-eyed) are called **homozygous** arrangements (the two genes are identical). Bb, which also produces brown eyes, is termed **heterozygous** (the two genes differ).

2. *The Law of Segregation*. Genes occur in pairs (termed **alleles** when they vary somewhat from one another). Each pair of genes resides on a pair of chromosomes (**Figure 1-3**).

Figure 1-3. A pair of chromosomes with corresponding gene alleles. In human **somatic cells** (any cell not a reproductive cell), one chromosome originated from the father, the other from the mother.

Humans have 23 pairs of chromosomes, totaling 46 chromosomes, which is termed the **diploid** number (**Figure 1-4**).

Figure 1-4. The 23 pairs of human chromosomes. Chromosome pair 23 may be either female (XX) or male (XY).

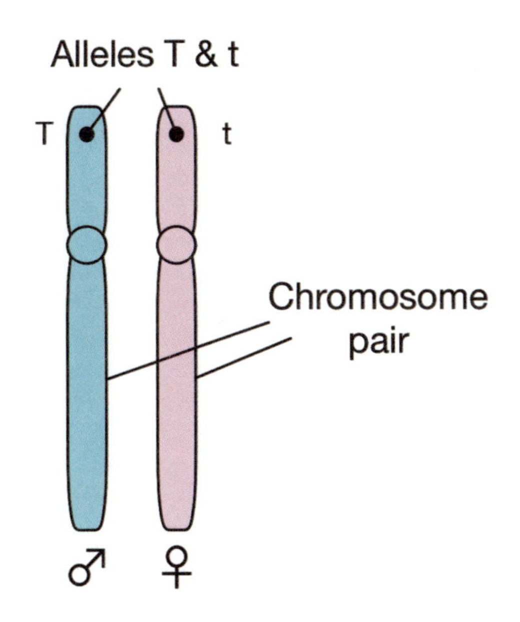

Fig. 1-3.

The two alleles of a pair segregate from one another when forming a sex cell (**gamete**), which contains 23 chromosomes (termed the **haploid** number). When the gametes (sperm in male and ovum in female) combine, the individual chromosomes come together as pairs, restoring the diploid number of 46, with one chromosome from the father and the other from the mother, for life.

3. *The Law of Independent Assortment*. In the classic Mendelian model, the segregation percentages of one pair of genes, as in **Figure 1-2**, occurs independent of the segregation that occurs for other genes on other chromosomes.

What makes a gene dominant or recessive? Is there something mystical about dominance or recessiveness? A defective gene for a disease is considered dominant if it produces something that is harmful to the person, or if it produces too little of a needed substance, which its corresponding allele cannot compensate for. The one defective gene alone causes the disease.

A recessive gene for a disease may similarly produce something that is harmful to the person or produce too little of something that is needed, but is compensated for by the normal gene. *Inborn errors of metabolism*, where there is a missing enzyme, typically are recessive disorders, since the amount of enzyme produced by the normal gene alone is enough to compensate for the decreased enzyme from the defective gene. Only when both genes are defective does the disorder become manifest. A gene is dominant when it alone can cause the respective phenotype. It is recessive when both genes are necessary to produce the phenotype.

Mendel's rules, though, have many exceptions:

- There can be many alleles in the community for the same gene, not just two. E.g. there are three versions of the allele for blood type: type A, type B, and type O. However, a given person can have only two of them, since chromosomes and their associated genes are paired. In some clinical conditions, there can be hundreds of different versions of mutations to a single gene, but a person can only have two at most.
- Mendelian inheritance refers only to single gene mutations, and there are indeed thousands of diseases that result from single gene mutations. Much of genetic inheritance, though, is more complex than that, involving problems with multiple genes on different chromosomes that act together to determine the phenotype. For instance, many genes in one person can determine the degree to which diabetes mellitus, hypertension, heart disease, autism, or obesity is manifest. One gene can affect the degree of expression of another

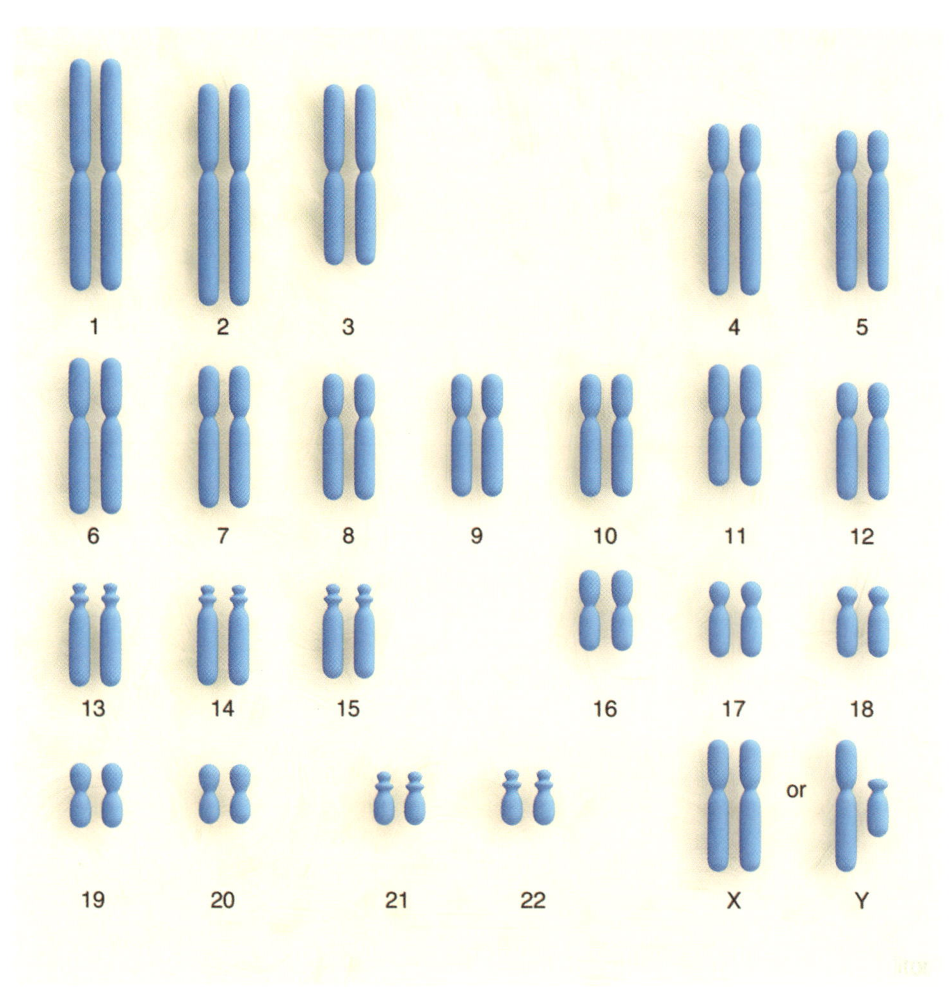

Fig. 1-4.

gene (e.g. in suppressing it) that may even lie on a different chromosome (termed **epistasis**).

- Some disease manifestations depend largely on environmental (**epigenetic**) influences, which affect gene expression with no mutation to the gene structure, but with an altered phenotype nonetheless. For instance, diet can affect the manifestations of diabetes, without changing the gene structure.
- Genes may not necessarily show strictly dominant or recessive traits, but can result in an in-between phenotype in the heterozygous state. Thus a patient may exhibit a full-blown disease in the homozygous dominant state, but a partial manifestation of the disease when heterozygous. For example, patients homozygous for the sickle cell gene demonstrate the full disease, but people who are heterozygous may demonstrate the disease only under certain environmental conditions (e.g. low pressure in an airplane, dehydration, intense physical activity). In some cases, it is not clear why certain people demonstrate only certain aspects of a syndrome while others are more fully affected. This may have to do with other differences in the genetic makeup between individuals, either other genes that influence the mutant gene, or differences in the environment that influence gene expression.

Looking more closely at Mendel's peas, some peas are normally smooth and some are wrinkled. The smooth variety (S) appears dominant over the recessive wrinkled variety (s), just as the tall and short qualities of the pea plants were dominant and recessive respectively. Peas are visibly smooth (the dominant trait) because they have a greater concentration of starch grains inside than does the wrinkled (recessive) variety. But if you microscopically examine the cut-open peas, the concentration of starch grains in the heterozygous smooth peas (Ss) lies in-between that of the homozygous dominant (SS) and homozygous recessive (ss) pea. Even though SS peas and Ss peas are both smooth to the naked eye, there is a hidden difference; the Ss pea has fewer starch grains than the SS pea. Thus, when we call a condition dominant or recessive, there may actually be hidden intermediate differences. (While this is a kind of "mixing," it is one whose elements can be sorted out in later generations.)

- Regarding the Law of Segregation, genes do not always neatly separate from one another in forming a sex cell, and a variety of chromosomal anomalies can arise, with altered numbers of chromosomes or copies of genes.
- The Law of Independent Assortment does not follow through when two different genes lie on the same chromosome, thereby linking them as if they were one, in which case they do not sort independently of one another.
- A striking exception to Mendel's rules is the finding that some genes have numerous identical copies that may even reside on unrelated chromosomes! How many of these copies are functional? This really throws a monkey wrench into the rules.
- Different genes with the same function can lie on different chromosomes ("One phenotype, multiple genes"). For instance, the connective tissue disease Ehlers-Danlos can arise from mutations in different genes on different chromosomes.
- Sometimes a mutation in a single gene can manifest as different phenotypes (**pleiotropy**) ("One gene, multiple phenotypes"). For instance, the gene mutation for phenylketonuria can cause intellectual disability as well as skin rashes.

What Exactly Is a Gene?

The specific definition of a gene that some people use is that of a DNA sequence that encodes for a protein. Some DNA sequences don't appear to encode for anything and are affectionately referred to as "junk" DNA but here we will expand on the definition to include a DNA sequence that encodes for anything useful. This includes certain RNAs that have different

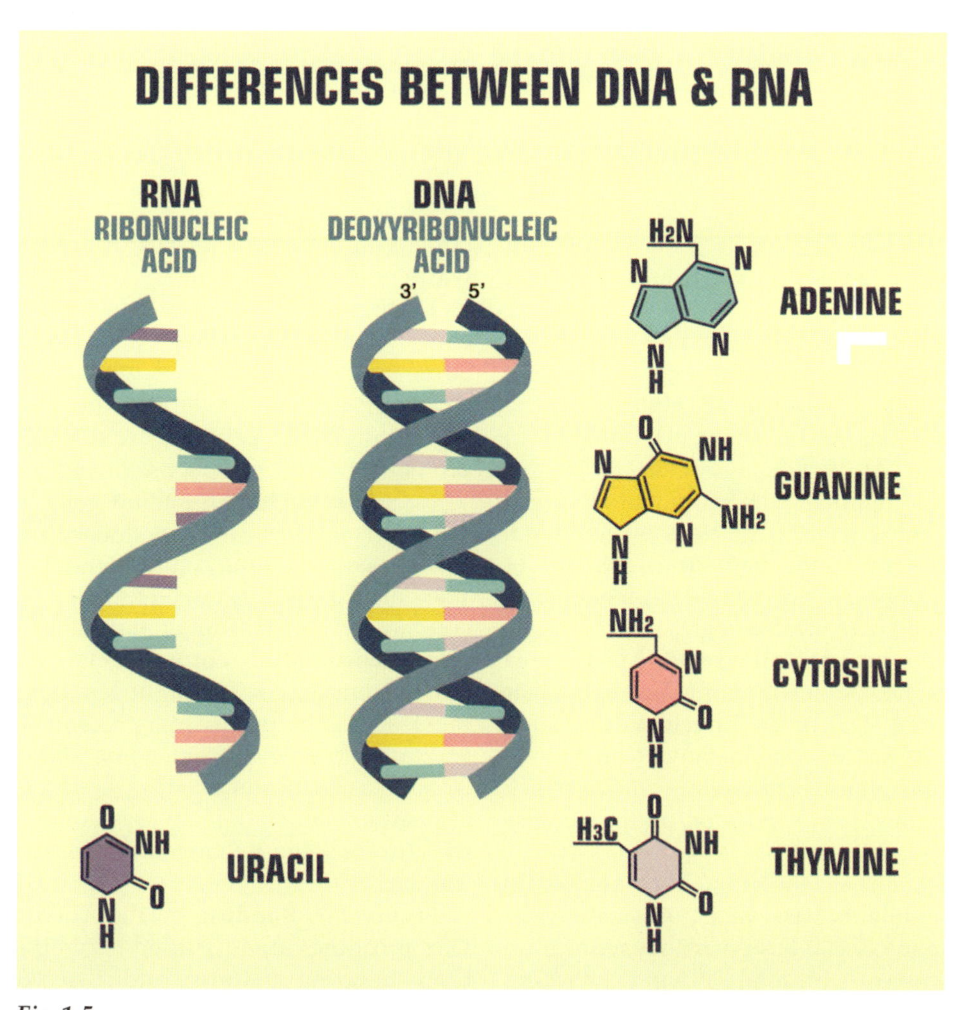

Fig. 1-5.

DNA STRUCTURE

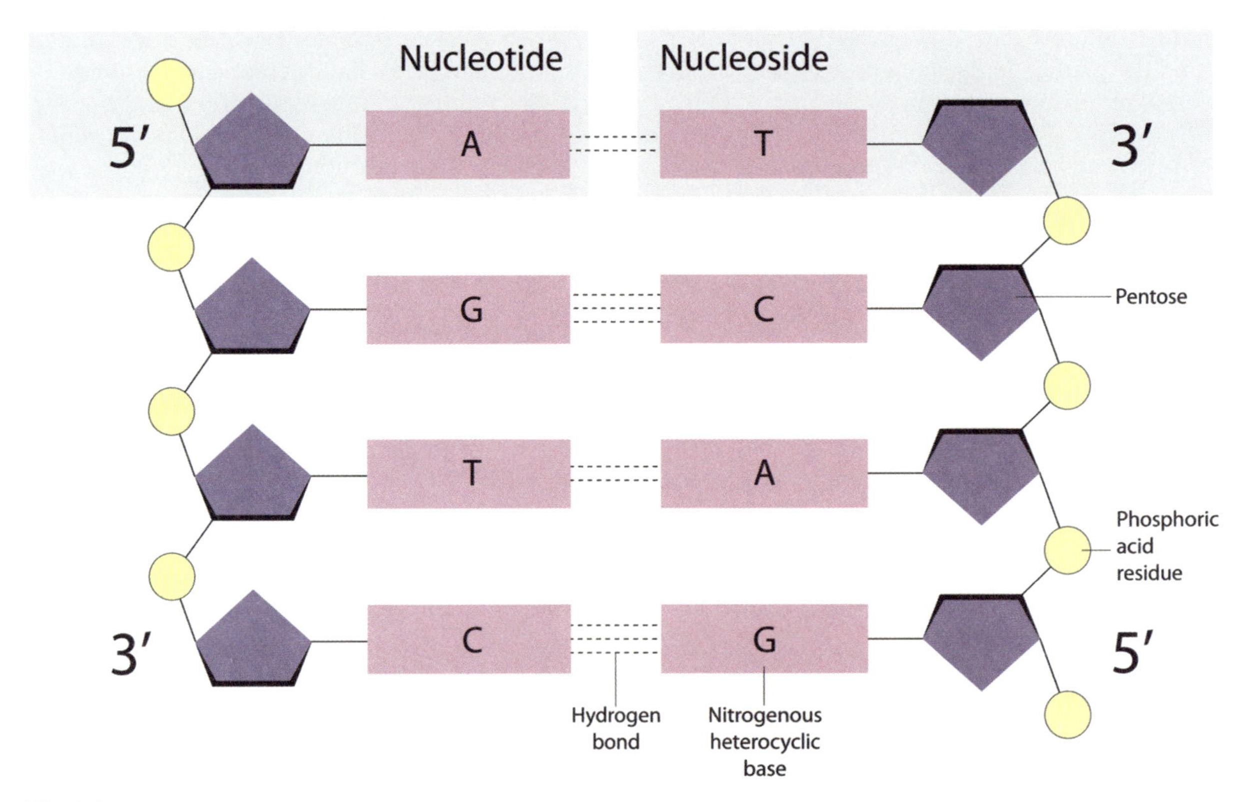

Fig. 1-6.

functions other than encoding for proteins. Only about 2% of DNA actually codes for protein. Much of the rest is a mystery.

DNA is a double-stranded helix, containing a winding ladder of 4 kinds of connecting nucleotide base pairs—*adenine*, *thymine*, *cytosine*, and *guanine* (**Figures 1-5** and **1-6**). Adenine (A) bonds with thymine (T); cytosine (C) bonds with guanine (G).

Figure 1-5. DNA differs from RNA in that DNA is a double-stranded helix, while RNA is single-stranded. RNA contains the nucleotide base uracil instead of thymine and also ribose, a 5-carbon sugar, while DNA contains the 5-carbon deoxyribose sugar.

Figure 1-6. DNA structure. A **nucleoside** (shaded area on upper right) is the combination of a 5-carbon sugar (pentose) and base (A, T, G, or C). A **nucleotide** (shaded area on upper left) is the combination of nucleoside and connecting phosphoric acid (yellow circle), which connects to either the 5′ or 3′ position on the pentose sugar ring.

The total DNA on all the human chromosomes contains about 3 billion base pairs. A single gene commonly has about 27,000 base pairs, sometimes up to 2 million. Stretched out, the total DNA in a cell would extend about 6 feet (~ 2 meters). All the DNA in one body would stretch out to a distance greater than the width of the solar system!

There are hundreds to thousands of genes on each chromosome. Humans have about 20,000-25,000 genes. Some chromosomes have more genes than others. Chromosome 1, the largest of the chromosomes, has about 8,000 genes. Chromosome 21 has about 225 genes.

Genes cannot be seen with regular light microscopy, but bands on chromosomes can be seen (**Figure 1-7**) by using special staining techniques. A band is not a gene. The dark bands are stained areas of **heterochromatin**, condensed chromosomal material. The light bands are **euchromatin**, chromosomal material that is not so condensed. The light bands appear to be the most active in producing protein. There are about 850 discernable chromosome bands, but about 20,000-25,000 protein-coding genes. A change in the appearance of a band can give some idea of where in the chromosome a change in a gene has occurred, but not which specific gene.

Figure 1-7. Banding on the 23 chromosome pairs. Chromosomes are identified by their size differences, their banding pattern, and the position of their centromeres. A **centromere** is the region of DNA on

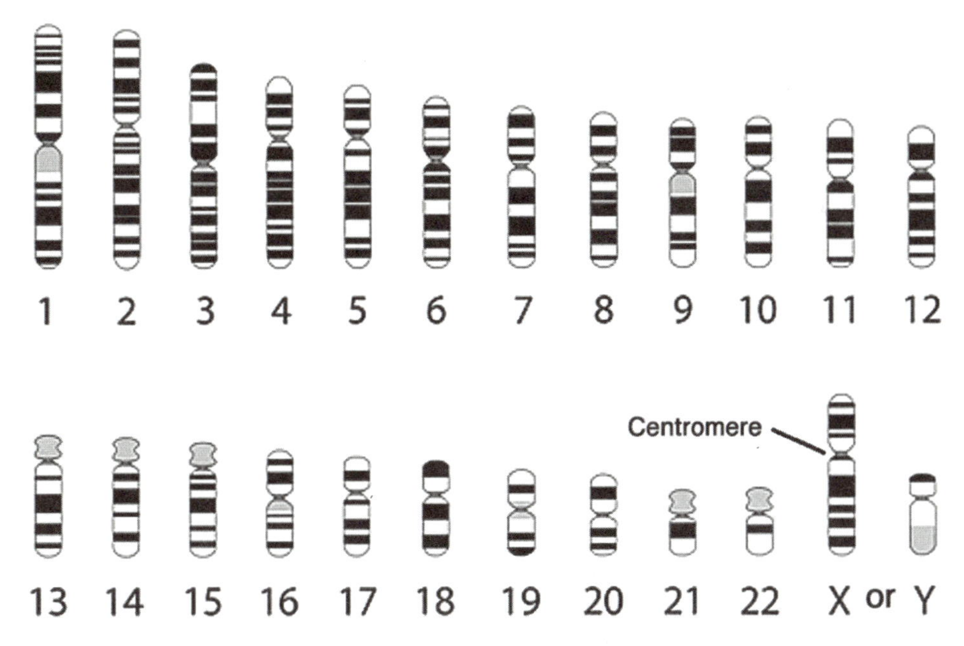

Fig. 1-7.

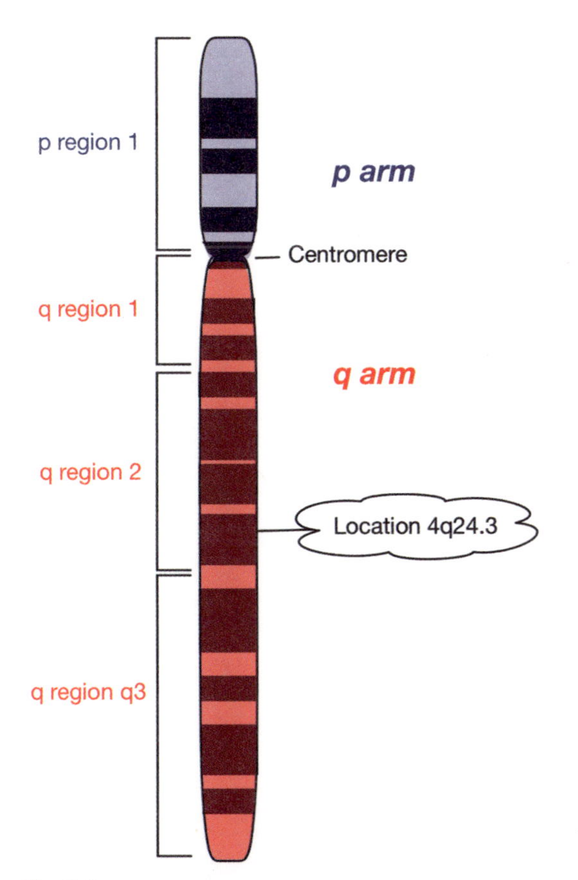

Fig. 1-8.

the chromosome that splits along with the rest of the chromosome during mitosis and is important for moving the split chromosomes into the two daughter cells during cell division (mitosis). Each chromosome has only one centromere.

Nomenclature

Locations on chromosomes are named counting outward from the centromere, "q" referring to the long arm of the chromosome in relation to the centromere, and "p" referring to the short arm (p is for petite). The first number is the number of the chromosome. For instance, location 4q24.3 (**Figure 1-8**, not drawn to scale) lies on the long (q) arm of chromosome 4, region 2, band 4 of region 2 (counting away from the centromere), sub band (subdivision) 3 of band 4. (A region is not a band; there may be only a few designated regions on a chromosome arm. And a band is not a gene, as a band may contain many genes.)

Today, with the sequencing of the genome, it is possible to use a more precise molecular terminology that points to the exact location of a particular nucleotide sequence on the chromosome of interest.

Genetic diseases have no set nomenclature. They can be named after:

1. The patient or family having the disorder (e.g. *Lou Gehrig disease*)

2. The biochemical defect (e.g. *phenylketonuria*)
3. The part of the body involved (e.g. *polycystic kidney disease*)
4. One or more of the major signs and symptoms (e.g. *familial hemiplegic migraine*)
5. The name of the doctor who discovered the disease (e.g. *Parkinson disease*)
6. The region of the world where the disease is particularly common (e.g. *Mediterranean fever*)

The naming of genes is more tricky, difficult to predict, and subject to change. They may be named using acronyms relating in some way to the gene's character or function. For example, BRCA1 = breast cancer gene 1; LDLRAP1 = Low-Density Lipoprotein Receptor Adaptor Protein 1 gene. They may also be named after a protein they produce (e.g. the p53 gene produces the tumor protein p53).

2

Mitosis And Meiosis

Mitosis is cell division; it is needed for the growth and differentiation of the body tissues from the earliest stages of development to the adult. Mitosis also functions to replace damaged cells, as in wound healing, and replenish cells that normally continue to divide and die throughout life (e.g. the cells of the intestinal lining, which are continually replaced). **Meiosis** is the specific form of cell division that results in the sex gametes (sperm and ovum), which, on fertilization, set in course the development of the individual.

Mitosis

Humans have 23 pairs of chromosomes, totaling 46. Before a cell undergoes mitosis, each chromosome replicates its DNA, forming two **chromatids** (**Figure 2-1**), which are attached together via a centromere. A chromatid is not a chromosome, but a chromosome that is dividing. If you want to know how many chromosomes a cell has, count the centromeres. In mitosis there are 46 chromosomes, 46 centromeres and 92 chromatids. The chromatids separate from one another, allowing each of the 2 daughter cells that result from a mitotic division to maintain the chromosome number at 46, termed the **diploid** (2n) number. The cells that divide after fertilization and throughout life are termed **somatic cells**, as opposed to **reproductive cells**, which are the cells that form the sex gametes by the process of Meiosis.

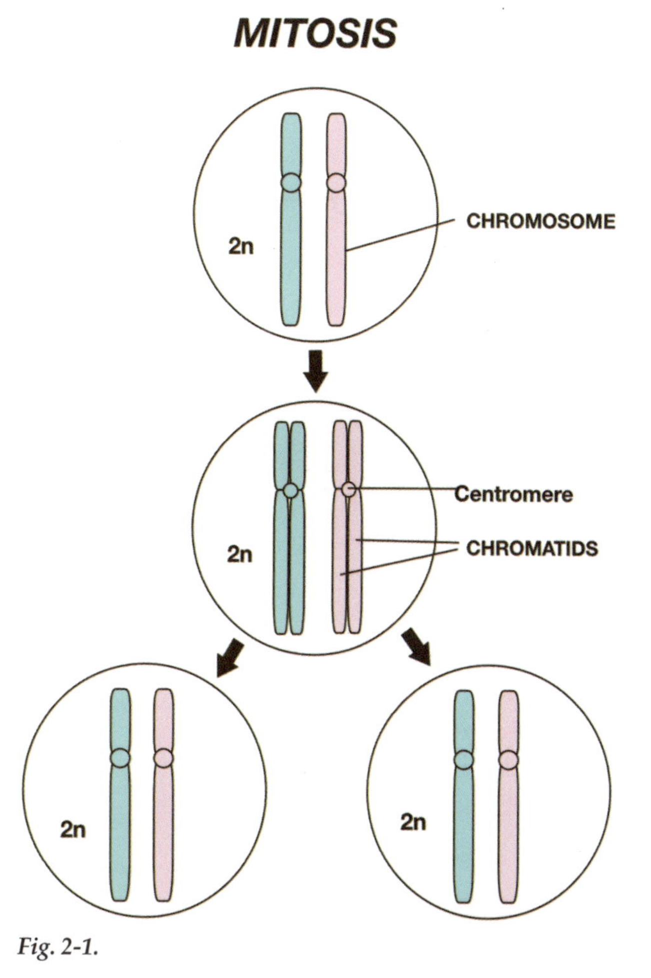

Fig. 2-1.

Meiosis

In Meiosis, which occurs in the testes and ovaries, the number of chromosomes in each reproductive cell is halved, forming sex gametes: sperm in males and ovum (or egg cell) in females, which contain only 23 chromosomes each, termed the **haploid** (n) number. When these gametes combine (fertilization), the resulting cell is called a **zygote**, with the diploid number of 46 chromosomes (2n) restored. The 23rd pair of chromosomes is the sex-linked pair, X and Y. A person with two X chromosomes (XX) is female, and a person with one X and one Y chromosome (XY) is male, with Y determining the male phenotype. The Y chromosome is much smaller than the X chromosome.

Figure 2-2. Meiosis in the male. Chromosome number is halved in Meiosis I, at which time **homologous** (part of the same pair) chromosomes exchange genetic material. In Meiosis II, which is an ordinary mitotic division, the individual chromatids separate into 4 different gametes.

Figure 2-3. Meiosis in the female. Division occurs as it does in the male, but with the formation of unused **polar bodies** (small cells that do not develop into ova and cannot be fertilized).

Meiosis has two stages of cell division—Meiosis I and Meiosis II, each of which has a distinct function (**Figures 2-2** and **2-3**). In Meiosis I, which occurs in both the testes and ovaries, the number of chromosomes is halved. So why do we need Meiosis II, if the chromosomes are already halved in Meiosis I, which is what we want? It seems that throughout all of life, the chromosomes themselves never have sex. Each pair of chromosomes that a person has consists of one chromosome from the person's father and the other from the person's mother. But they never touch each other to consummate their partnership. In Meiosis I, however, the chromosomes in a pair contact each other, exchanging chromatid segments (**synapsis**) (**Figure 2-4**). Then, out of exhaustion from the process, when the cell divides, the chromatids of each chromosome are too tired to separate from the centromere; so each chromosome of the pair (each chromosome containing two chromatids) moves separately into 2 daughter cells (one of which disappears in the female as a **polar body**) (**Figure 2-3**), thereby halving the number of chromosomes in a cell by the end of Meiosis I. Note, however that chromatids in the daughter cell now contain one part from the person's father and another part from the person's mother. This ensures that the genes of father and mother are shuffled, so that part of the inheritance will be from the father and part from the mother, providing genetic variety. The crossing may occur in one or more parts of the chromatid.

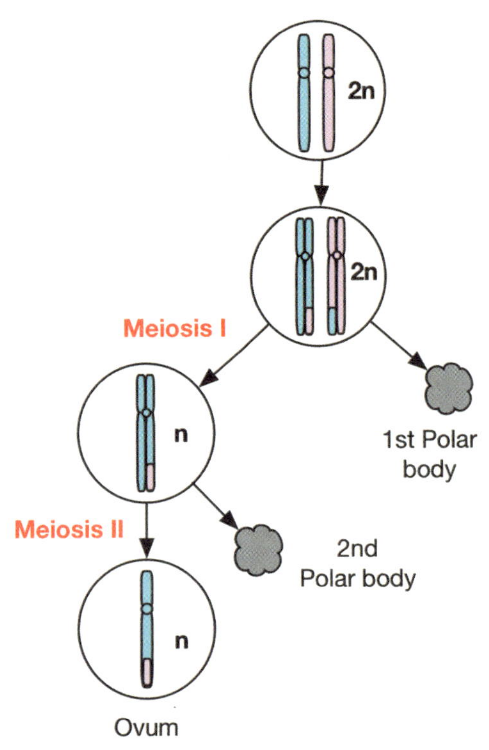

Fig. 2-3.

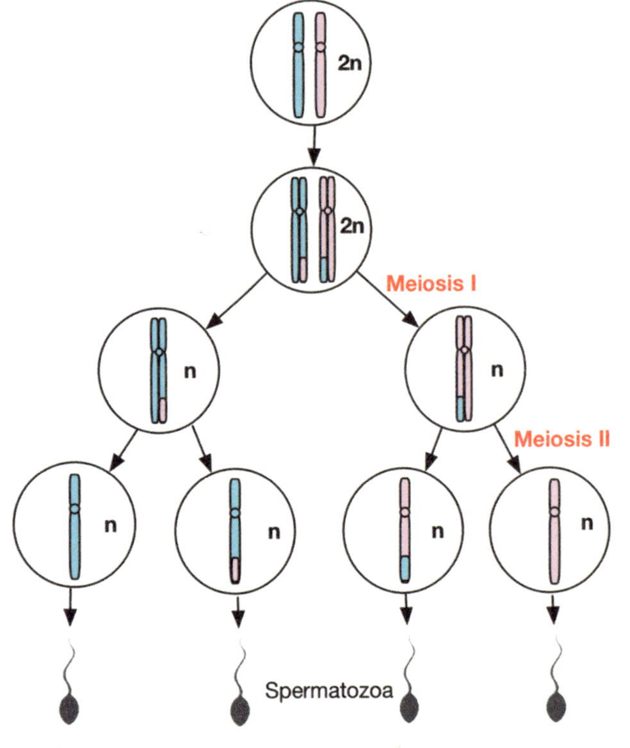

Fig. 2-2.

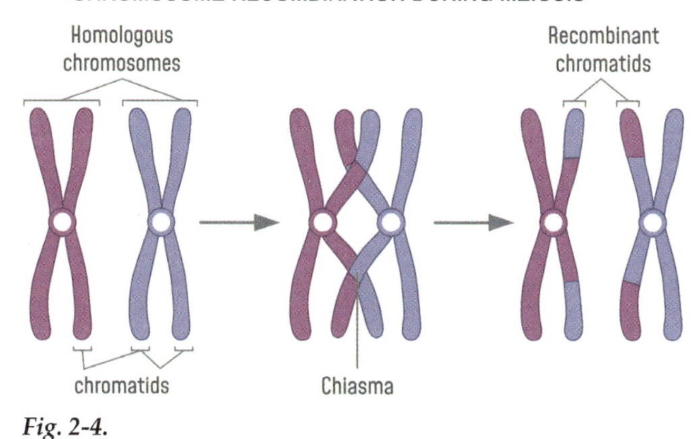

Fig. 2-4.

Figure 2-4. Chromosome recombination during Meiosis I. Crossed exchange of chromosome pieces may occur in one or more chromosomal locations.

In Meiosis II (which occurs in the testes, and in the Fallopian tube only after fertilization), the cell (now haploid) undergoes a regular mitosis; the chromatids in the cell separate from one another during this division, forming two haploid cells (**gametes**), which differ from one another in their chromosome gene content (**Figures 2-2** and **2-3**). In the male, the total number of gametes formed is four, but in the female, only one gamete persists, due to the disappearance of polar bodies. When a sperm cell fuses with the egg, forming the zygote, this restores the diploid number of 46. One chromosome of each pair is still from the father and the other is still from the mother, for life.

The exact percentage of sorting of the different kinds of genes in the gametes depends on how near to one another the genes are on the chromosome. If they are very close, they stick together as one (**linkage disequilibrium**). If they are far apart, they may sort to increasing degrees in Meiosis I, according to how far apart they are (**Figures 2-5** and **2-6**).

Figures 2-5 and **2-6**. Linkage equilibrium and disequilibrium. Genes closer together are less likely to separate during Meiosis I.

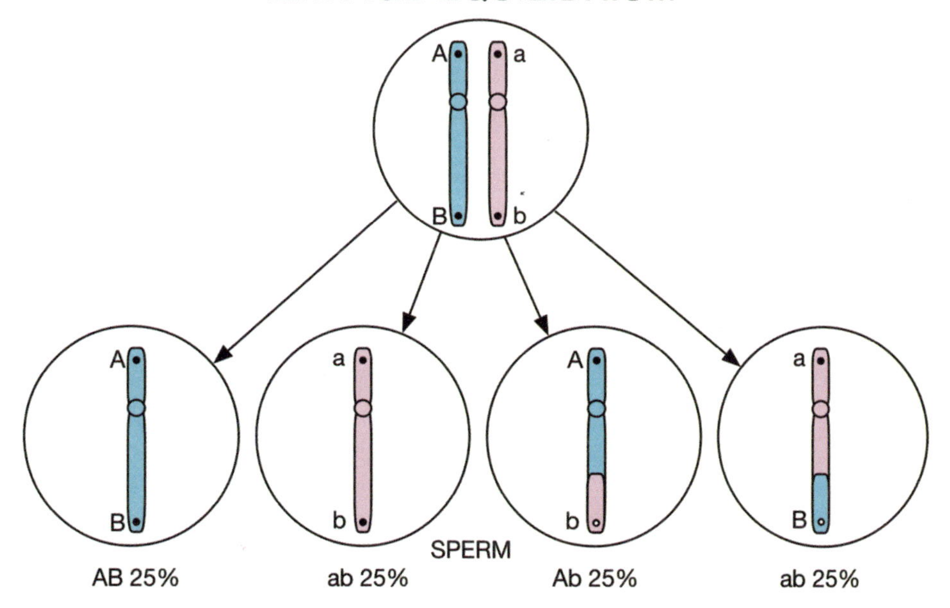

Fig. 2-5.

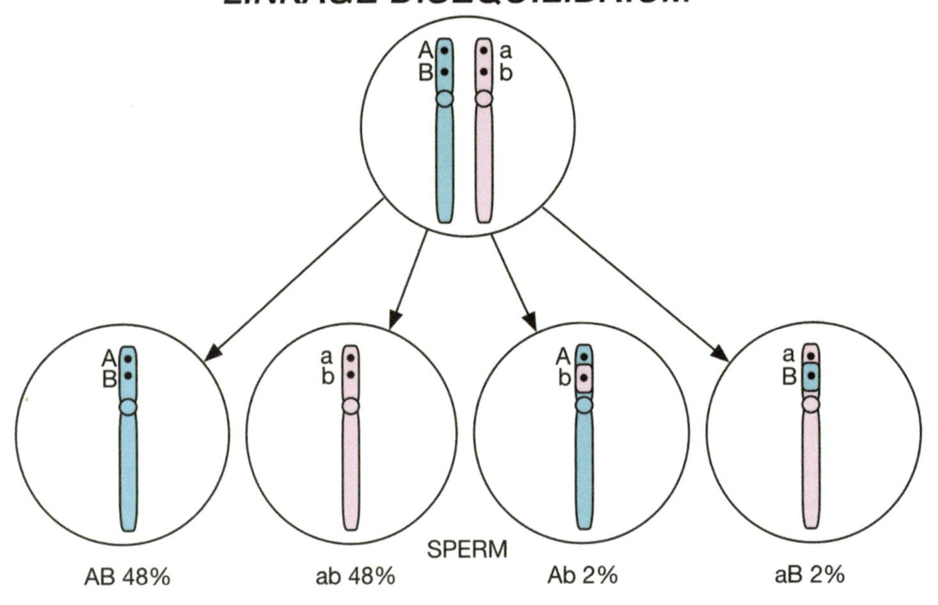
LINKAGE DISEQUILIBRIUM
A
B
a
b
A
B
a
b
A
b
a
B
SPERM
AB 48%
ab 48%
Ab 2%
aB 2%

Fig. 2-6.

3

DNA Replication, Transcription, and Translation

The key events in the formation of DNA, RNA, and protein are **Replication** (of DNA), **Transcription** (of DNA into RNA), and **Translation** (of RNA into proteins). The main aim of all this is the production of proteins, which are important actors in virtually all cell processes. A **protein** is a large molecule containing one or more long chains of **polypeptides**. A **polypeptide** is a chain of many amino acids. A **peptide** is two or more amino acids joined together. For simplicity of discussion, this book will refer to all three as simply "proteins." **Figure 3-1** shows the 20 amino acids that are used to construct proteins during translation.

Figure 3-1. The 20 amino acids that string together to form proteins. (Actually, there are 22, if you include *hydroxyproline* and *hydroxylysine,* but these two do not participate in the original formation of the protein; they are modifications on the protein once the protein is formed. They play an important role in the structure of collagen.)

Genetic diseases can arise from DNA mutation or by interference with any of the steps in replication, transcription, or translation, or in processing the protein once it is formed. It is important to understand these steps, since therapy may depend on where the problem lies.

Replication of DNA

DNA replication in mitosis requires a number of actors (**Figure 3-2**):

Figure 3-2. Steps in the replication of DNA. The two DNA strands first have to separate. Each strand will be used as a template for replicating new DNA, in the directions shown by the arrows.

- A **helicase** enzyme separates the two strands in the DNA double helix to allow room for new nucleotide bases to get in and bind to their complementary bases to form new DNA, using each strand as a template.
- A **topoisomerase** enzyme uncoils the separating double-stranded DNA.
- A **primer** (**Figure 3-3**) is needed to start off replication. The primer (a short RNA sequence, synthesized by a **primase** enzyme) attaches to each template strand and is the starting point for the addition of nucleotides to form the new DNA strand. The RNA primer is later removed and replaced by DNA nucleotides.
- A **DNA polymerase** enzyme polymerizes (joins together) the nucleotides to form a new DNA strand. During DNA synthesis, the polymerase, avid foreman that it is, also "proofreads" to check that the job of adding the appropriate base pairs is going smoothly, correcting any errors in the replication—removing, or refusing to allow the addition of, inappropriate nucleotides.

Figure 3-3. Sequence of formation of the replicating DNA strands (red arrows). The two strands of the

glycine (Gly, G) L-alanine (Ala, A) L-valine (Val, V) L-leucine (Leu, L) L-isoleucine (Ile, I)

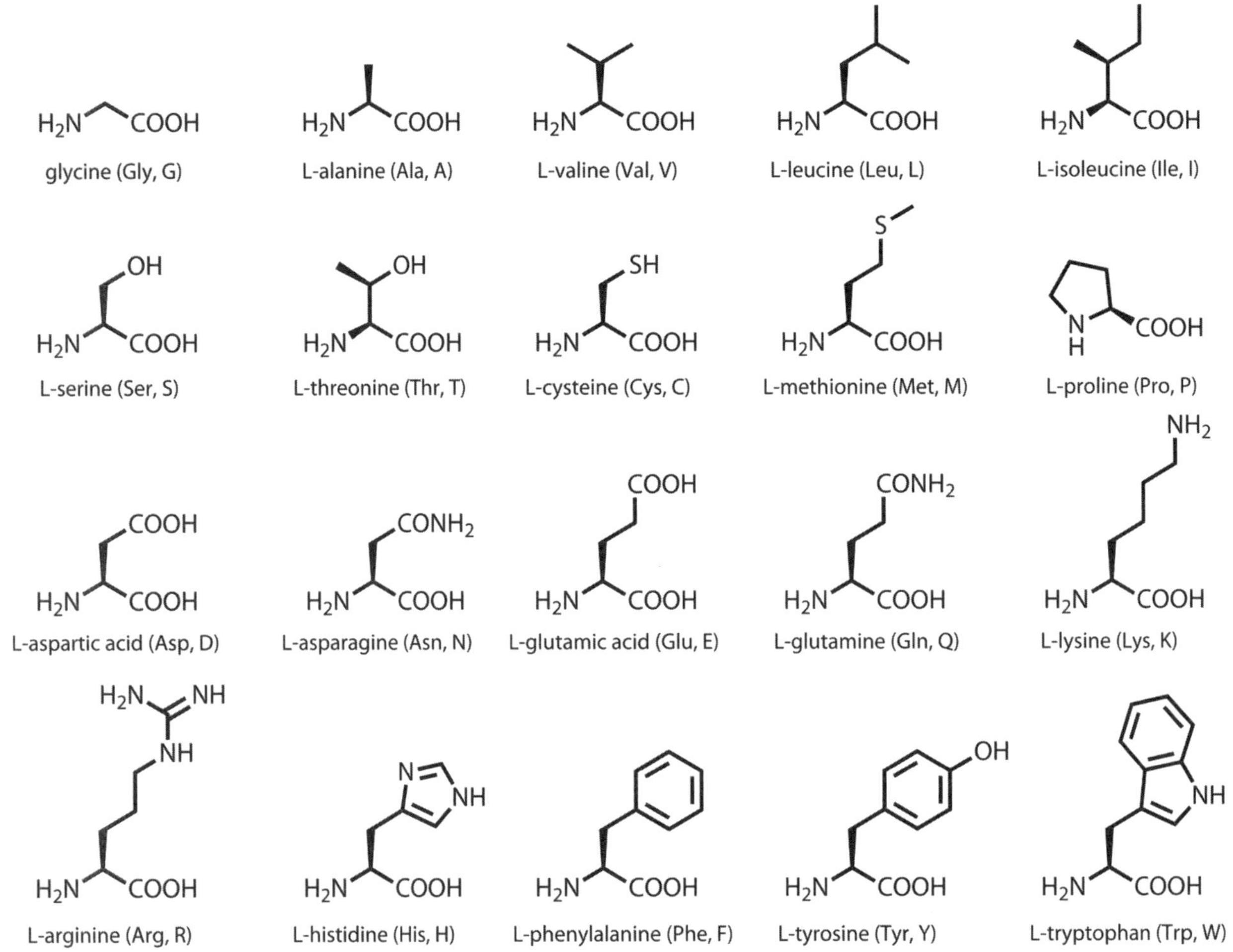

L-serine (Ser, S) L-threonine (Thr, T) L-cysteine (Cys, C) L-methionine (Met, M) L-proline (Pro, P)

L-aspartic acid (Asp, D) L-asparagine (Asn, N) L-glutamic acid (Glu, E) L-glutamine (Gln, Q) L-lysine (Lys, K)

L-arginine (Arg, R) L-histidine (His, H) L-phenylalanine (Phe, F) L-tyrosine (Tyr, Y) L-tryptophan (Trp, W)

Fig. 3-1.

normal DNA helix are arranged in an antiparallel manner, such that the 5′ to 3′ pentose sugar ring chemical bond sequence (see **Figure 1-6**) of one strand is in line with the 3′ to 5′ sequence of the other strand. Nucleotides can only be added in the 3′ to 5′ direction of the uncoiling DNA strand. This is easy for one of the DNA template strands, which happens to run in the 3′ to 5′ direction. It's complementary strand, though, runs in the 5′ to 3′ direction, so the addition of nucleotides to the latter strand has to be done backward, awkwardly in segments, called **Okazaki fragments** as the DNA unwinds.

- A **DNA ligase enzyme** glues together the newly formed Okazaki fragments of DNA (**Figure 3-3**).
- After replication, **endonuclease** and **exonuclease** enzyme repair teams come in to remove any defective segment of DNA, whether an erroneous deletion or insertion during DNA replication that was overlooked. The endonuclease does this removal in the middle of the DNA chain, while the exonuclease does the removal at the end of the chain.
- Another **DNA polymerase** enzyme (there are at least 14 different kinds) steps in to fix the damaged DNA (**mismatch repair**). There thus is more than one foreman for replication. One DNA polymerase synthesizes a strand of DNA during replication and proofreads it, correcting any error during replication; another one repairs DNA after replication has taken place.

Why mention all these steps in a medical book? Are they clinically relevant? They are, because different diseases can affect any of the steps in the replication process, or for that matter any of the steps in the transcription of DNA into RNA, or RNA into protein, or modification of the proteins that are produced at the end. The protein enzymes involved in these actions are themselves products of DNA, so DNA mutations can involve any of them, leading to a defect in their function. There are thus diseases of DNA repair, diseases

DNA replication

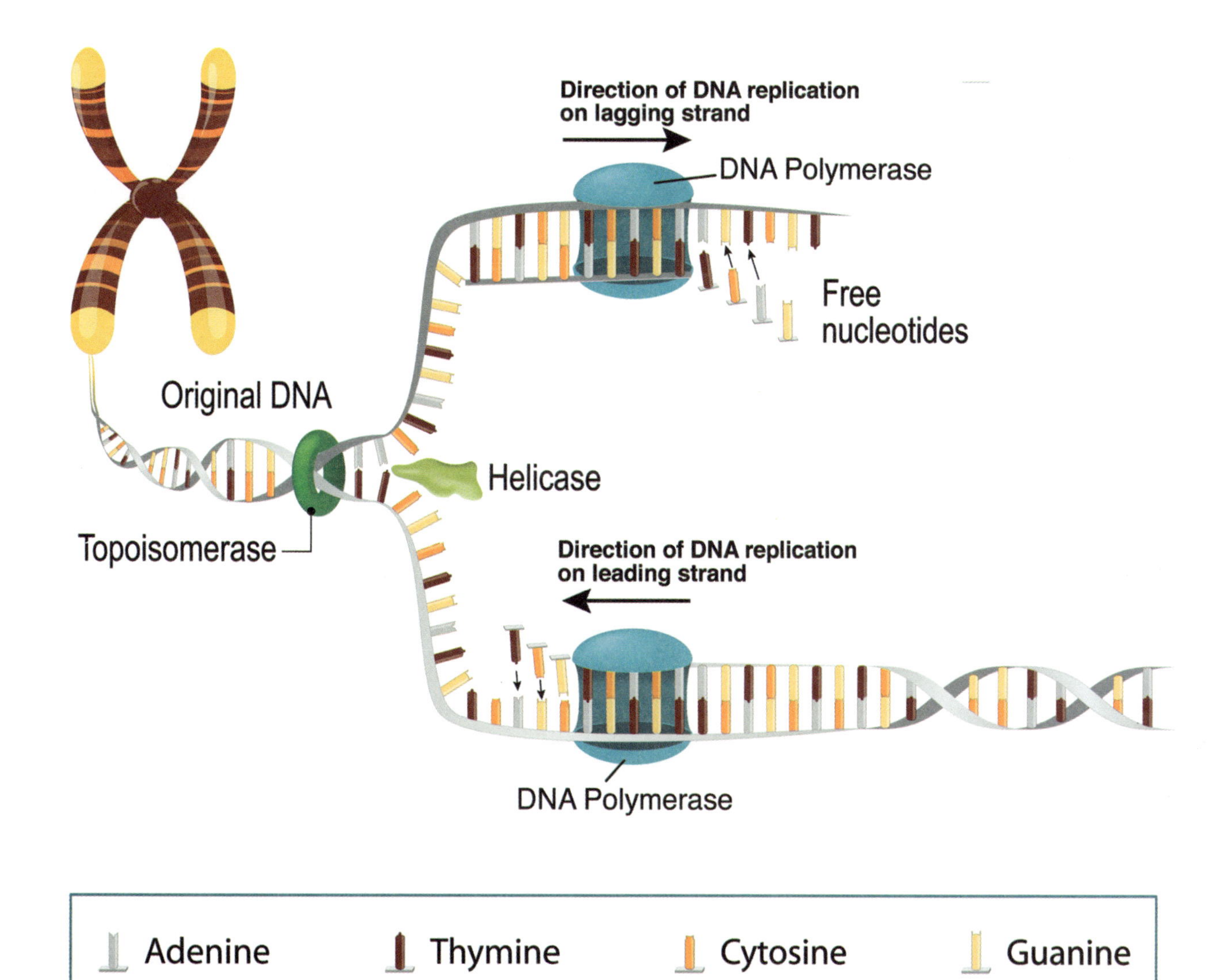

Fig. 3-2.

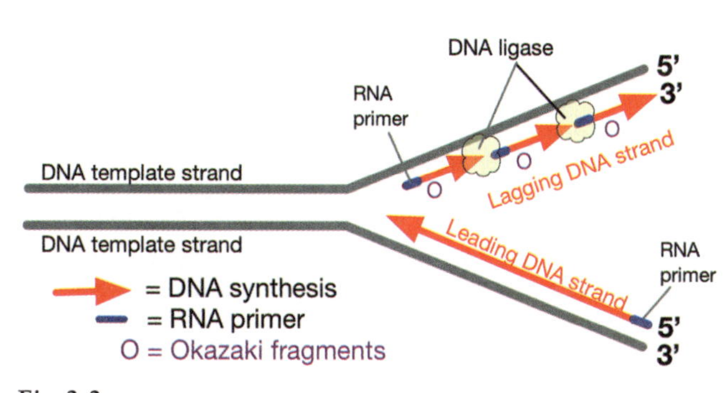

Fig. 3-3.

of unwinding the DNA, of DNA ligase, and of the replication fork (**Figure 3-4**).

Environmental influences (e.g. diet, temperature, toxic agents, infection) also play a prominent role in modifying the functions of all these components of the replication process.

Transposons

Amid all the goings-on of DNA replication, there is the phenomenon of "jumping genes." Yes, there are segments of genes, termed **transposons**, that can jump to a different site on a chromosome or even

FIGURE 3-4. EXAMPLES OF DISEASES OF DNA REPLICATION AND REPAIR (See the Appendix of Diseases for expanded detail on each disease.)
• Autism (multifactorial) • Aicardi-Goutieres syndrome (exonuclease) • Ataxia telangiectasia (helicase; DNA ligase; phosphorylation during DNA repair; exonuclease) • Baller-Gerold syndrome (helicase) • Bloom syndrome (helicase; DNA ligase) • Breast/ovarian/prostate/pancreatic cancer/melanoma (DNA repair) • Cerebro-oculo-facial skeletal syndrome (helicase) • Cockayne syndrome (excision repair; helicase) • Fanconi anemia (replication fork; DNA ligase; exonuclease; helicase) • Hutchinson-Gillford progeria syndrome (repair of double-stranded breaks) • Li-Fraumeni syndrome (DNA repair) • Ligase 4 syndrome (DNA ligase) • Lynch syndrome (mismatch repair) • Nijmegen breakage syndrome (exonuclease) • Rapadilino syndrome (helicase) • Roberts syndrome (helicase) • Rothmund-Thomson syndrome (helicase; DNA repair) • Scleroderma (topoisomerase) • Spinocerebellar ataxia with axonal neuropathy (topoisomerase) • Trichothiodystrophy (excision repair; helicase) • Warsaw breakage syndrome (helicase) • Werner syndrome (helicase; exonuclease) • Xeroderma pigmentosum (excision repair; DNA polymerase; DNA ligase; helicase)

onto a different unrelated chromosome. A whole gene commonly is about 27,000 base pairs long, but may range from a few thousand to over 2 million base pairs. Transposons are commonly several hundred to about 21,000 base pairs long. There are many thousands of transposons in the genome. Transposons can create mutations by moving into the middle of other genes. Mutations from transposons can contribute positively to genetic variation. However, they can also be harmful (being associated with cancer, aging, autoimmunity, neurodegenerative and other diseases) when they interrupt the middle of genes.

Transposons often duplicate the original genetic material. **Class II transposons** are those that can jump directly from one portion of the genome to another. They use an enzyme called a **transposase** (well named) to facilitate the jumping process. Sometimes Class II transposons leave a copy of themselves behind in the original location (a "copy and paste" mechanism). Or, they may not leave a copy behind (a "cut and paste" mechanism).

Class I transposons (also called **retrotransposons**) move indirectly. They are first transcribed into RNA, which then makes a copy of its corresponding DNA, which is then inserted elsewhere in the genome. Retrotransposons always leave a copy of DNA behind. They use a **reverse transcriptase** enzyme (one that can change RNA back into its complimentary DNA), the sort used by RNA retroviruses, like HIV, to insert themselves into host DNA. There may be an evolutionary relationship between retrotransposons and retroviruses.

LINEs (Long Interspersed Nucleotide Elements) and **SINEs** (Short Interspersed Nucleotide Elements) are Class I transposons (retrotransposons). LINEs consist of about 7,000 base pairs, while SINEs are about 100-700 base pairs long. LINEs and SINEs make up a large part of the genome and are scattered about the genome rather than being lined up in tandem. **Alu elements** (about 300 base pairs long) are the most abundant

FIGURE 3-5. SOME DISEASES LINKED TO TRANSPOSONS
• Alport syndrome • Alzheimer disease • Breast/ovarian/colon/lung/gastric cancers • Chorioretinal degeneration • Diabetes mellitus Type II • Ewing sarcoma • Familial hypercholesterolemia • Hemophilia A and B • Leigh syndrome • Leukemia • Mucopolysaccharidosis VII • Neurofibromatosis • Porphyria

SINE transposons; there are over a million copies of them, comprising about 10% of the genome. While Alu elements do not encode for protein, they are believed to play a role in modifying gene expression. **Figure 3-5** lists some human diseases that have been linked to transposons.

The copy and paste mechanism has in evolution left a lot of duplicated DNA in the genome. It is unclear presently how much of this is "junk" DNA with no meaningful purpose and how much has a significant function. In some cases, the number of repeats is correlated with the manifestation of various diseases, e.g. several cancers, due to mutagenesis, when transposons insert into and interrupt genes involved in regulating gene activity.

Tandem Repeats

Tandem repeats are repeats in the DNA sequence that are arranged in order next to one another. There are hundreds of thousands of them in the genome. They include **microsatellites** (consisting of only 2-6 base pairs) and **minisatellites** (larger, consisting of about 10-100 base pairs), either of which may repeat about 5-50 times. Tandem repeats can occur anywhere in a chromosome; many of them occur commonly in centromeres and telomeres.

Telomeres are regions of repeat nucleotide sequences (TTAGGG) at the ends of chromosomes. The repeats are believed to protect the chromosome from fusing with nearby chromosomes, which would result in abnormal chromosomal function. Normally, telomeres shorten with each cell division, leading to aging and **apoptosis** (programmed cell death). Many diseases related to aging involve the shortening of telomeres. The enzyme **telomerase** helps restore the telomeres and protects against their shortening. Telomerase elongates chromosomes by adding TTAGGG sequences to the ends of chromosomes. When cells do not use telomerase, they age, and the body ages. Cancer cells maintain their telomere length and don't die. Telomeres thus may play a role in aging and cancer. Mutations to the gene that encodes for telomerase have been associated with a number of diseases, including forms of *aplastic anemia* and *fibrosis of the lungs and liver*; shortened telomeres may also be a risk factor for developing cardiovascular disease.

Centromeres contain a large series of head-to-tail tandem repeats of non-protein coding DNA (**satellite DNA**), which is subject to mutation and epigenetic (non-genetic) influence. Disruption of centromere function can cause abnormal segregation of chromosomes during cell division, commonly seen in cancer cells. Centromere dysfunction is also correlated with certain autoimmune diseases (e.g. *scleroderma, lupus, rheumatoid arthritis, Sjogren syndrome*) due to the presence of *anticentromere antibodies*.

FIGURE 3-6. SOME DISEASES WITH GENETIC ANTICIPATION (NERVOUS SYSTEM GENERALLY)

- Amyotrophic lateral sclerosis
- Behcet disease
- Crohn disease
- Dentatorubral-pallidolusian atrophy
- Dyskeratosis congenita
- Familial breast and ovarian cancer
- Fragile X syndrome
- Friedreich ataxia
- Hereditary nonpolyposis colorectal cancer (Lynch syndrome)
- Huntington disease
- Li-Fraumeni syndrome
- Myotonic dystrophy
- Ovarian insufficiency
- Spinocerebellar ataxias

Tandem **trinucleotide repeats** (e.g. CCG, CAG) in DNA or untranslating regions of RNA have frequent mutations, and have been implicated in a variety of cancers and neurological and developmental disorders. The repeats may result in a defective protein, alter gene expression, or create a harmful RNA. In some diseases, the higher the number of duplications, the worse is the phenotypic defect. Diseases due to tandem repeats often show **genetic anticipation** (**Figure 3-6**), where the disease worsens with successive generations or arises at an earlier age. The worsening is correlated with an increasing number of repeats with each new generation.

Most repetitive DNA sequences do not code for proteins. They may be transcribed into RNA, though, in which case the RNA has a different function than making proteins, such as interacting with **chromatin** (a combination of DNA and histone protein) to alter gene expression.

Transcription of DNA into RNA

Transcription is the generation of RNA from DNA. In order for proteins to be synthesized, DNA must first generate RNA, which then must be translated into protein. RNA differs from DNA in being a single strand and having uracil in place of thymine in its base sequence; also DNA has deoxyribose as a sugar, while RNA has ribose.

RNA polymerase strings together nucleotides on the DNA template to form a complementary RNA copy (**Figure 3-7**) of a DNA gene sequence.

Figure 3-7. Transcription of DNA to RNA and translation to protein. mRNA = messenger RNA; tRNA = transfer RNA.

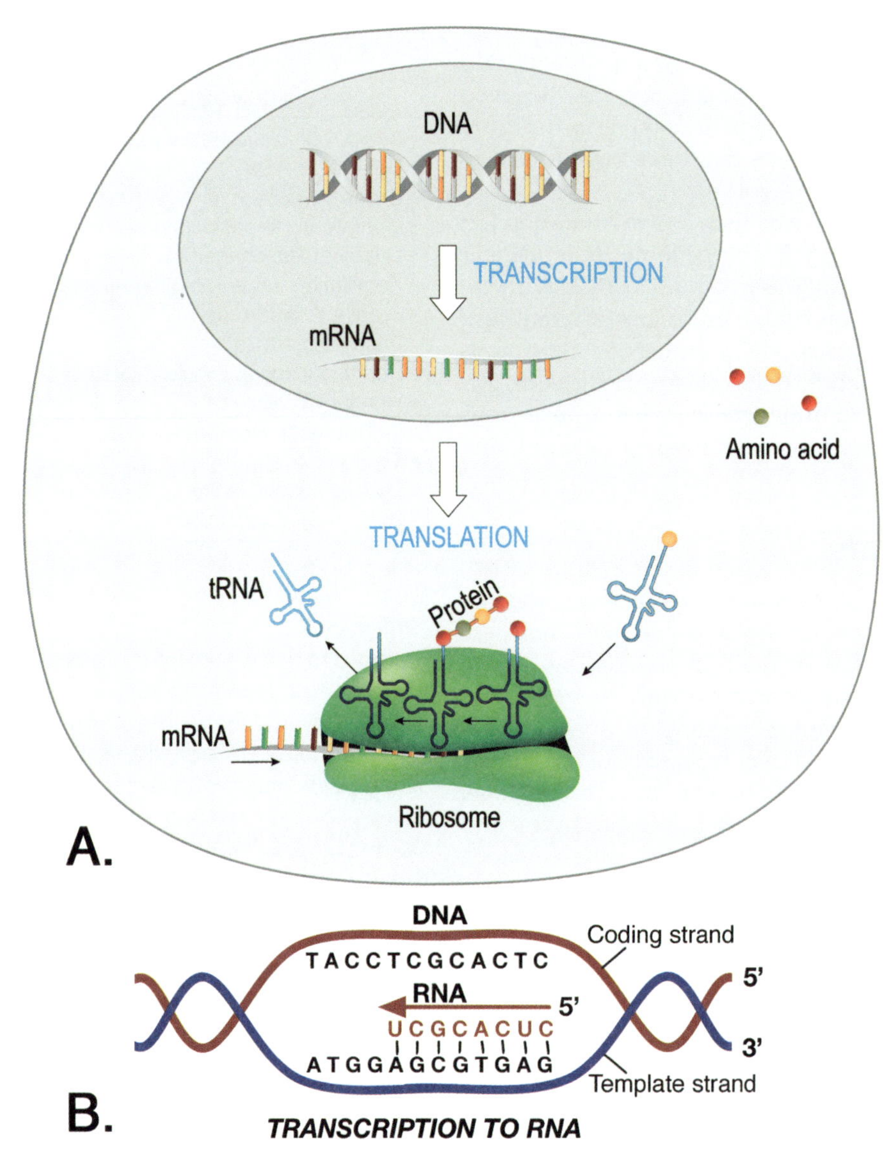

Fig. 3-7

Three general types of RNA that are involved in the pathway of protein production result from transcription (**Figure 3-7**):

- **Messenger RNA (mRNA)** carries the primary coded message of the specific protein to be synthesized on the ribosome, i.e. the specific sequence in which the amino acids are to be linked.
- **Transfer RNA (tRNA)** carries individual specific amino acids (one amino acid per transfer RNA molecule) to be lined up and linked up on the ribosomes under the direction of messenger RNA.
- **Ribosomal RNA (rRNA).** The nucleolus, which lies in the nucleus of the cell, is a factory for producing the cytoplasmic ribosomes. The nucleolus creates ribosomal subunits out of proteins and ribosomal RNA. These subunits then enter the cell cytoplasm, where the subunits combine to form complete ribosomes. The ribosomes help link the amino acids together to form proteins. Some ribosomal RNA, called **ribozymes**, are enzymes, catalysts for peptide bond formation.

DNA is responsible for the production of all proteins in the cells. DNA transcribes to form messenger RNA (mRNA), which carries the protein code out of the nucleus. DNA also transcribes to at least four other kinds of RNAs that do not in themselves encode for proteins: **transfer RNA**, **ribosomal RNA (rRNA)**, **small nuclear RNA** (**snRNA**, needed for splicing in the formation of mRNA – **Figure 3-9**), and **microRNA** (**miRNA**, which decreases gene expression). The latter two are discussed later in this chapter.

What marks the borders of a gene? **Promoters** and **terminators** are nucleotide sequences in DNA that mark the beginning and end of a gene, indicating the spots where RNA transcription should begin and end. (**Figure 3-9**). RNA polymerase binds to the promoter and starts the transcription of RNA, which stops on reaching the terminator, resulting in an RNA segment of fixed length. The RNA molecule may then be modified by adding or detracting from the RNA chain.

While a **primer** (a short RNA sequence), as mentioned previously in this chapter, is necessary to start *DNA replication* at the beginning of the DNA strand, a **promoter** (a region of DNA) marks the starting place on the gene for *RNA transcription* from DNA (**Figure 3-9**). A mutation in a promoter area of a gene can adversely affect the transcription of DNA into RNA and hence the production of the given protein (**Figure 3-8**).

Also, alterations in the methylation of promoters, without a direct mutational change in the promoter nucleotide sequence, can cause disease. **Methylation** (adding of methyl groups) of a promoter can inhibit a gene from expressing itself, while **demethylation** (removal of methyl groups) can cause the gene to overexpress. This has important implications in the development of cancer. For instance, methylation may silence a promoter in a gene responsible for DNA repair or tumor suppression, and hypomethylation may render a gene overactive in enhancing cell division.

Since DNA is double-stranded, which strand of DNA is used to generate the RNA, which is single-stranded? While both DNA strands are used for replication, only one strand is used in transcription, as a template for making RNA. There is a DNA **coding** strand that has the coding sequence for the future protein. There is also a corresponding **template** (“**sense**” template) DNA strand, which has a string of nucleotides complementary to those on the coding strand and is used as a template for creating the RNA (**Figure 3-7**). RNA can only follow the 3′ to 5′ direction on the DNA molecule, so it forms on the template strand only, but ends up duplicating the complementary nucleotide sequence on the coding strand (except for the substitution of uracil for DNA’s thymine).

FIGURE 3-8. SOME DISEASES OF PROMOTERS
• Asthma
• Bernard-Soulier syndrome
• Beta-thalassemia
• Charcot-Marie-Tooth disease
• Congenital erythropoietic porphyria
• Familial hypercholesterolemia
• Hemophilia
• Persistence of fetal hemoglobin
• Ornithine transcarbamylase deficiency
• Progressive myoclonus epilepsy
• Pyruvate kinase deficiency
• Rubinstein-Taybi syndrome

Introns

An **intron** is a special DNA sequence within a gene that does not code for proteins, but divides each gene into segments, called **exons**, which are the protein encoding parts of the gene. Introns have an important function. When RNA is transcribed from DNA, the introns are included in the transcription, but are later spliced out (**Figure 3-9**). Splicing allows exon segments to join in different combinations, providing for the generation of numerous different proteins from a single gene.

Figure 3-9. Intron splicing. RNA transcription extends from the promoter to the terminator end of a gene to form pre-mRNA. The introns in the pre-mRNA are spliced out to form the mature mRNA. The exons may rejoin in different combinations to form different mRNAs and hence different proteins. Sometimes exons are removed or introns retained.

Figure 3-10. Overview of transcription, splicing, translation, and protein folding. The **molecular cap** (C) and **poly-adenine tail** (AAA) endings on mRNA contribute to its stability. This is important because the longer the RNA hangs around, the longer it can continue translating into protein. What is important is not only the kind of protein that the RNA generates but how much and for how long. Stability in part depends on the length of the polyA tail, which may differ in different mRNAs.

How can so many millions of proteins arise from only 20-25,000 genes? Using a deck of cards analogy, although there are only 52 cards in the deck, there are millions of combinations of cards that can be formed from shuffling the deck. In the case of DNA, the intron interruptions in the DNA nucleotide sequence of a gene get spliced out of the RNA, so that the remaining segments of RNA can join together in different combinations. These combinations are a significant reason why so many types of proteins can arise from relatively few genes.

RNA itself can be edited by deletion, insertion, or substitution of nucleotides in the RNA nucleotide sequence (**RNA editing**), resulting in an even greater repertoire of proteins. This RNA editing is apart from the RNA modifications made through intron splicing, the molecular cap and poly-adenine tail (**Figures 3-9** and **3-10**).

In addition, the different proteins that form in the translation of RNA to protein can be further varied by being shortened or lengthened, or by combining with other proteins or other kinds of groups to form new

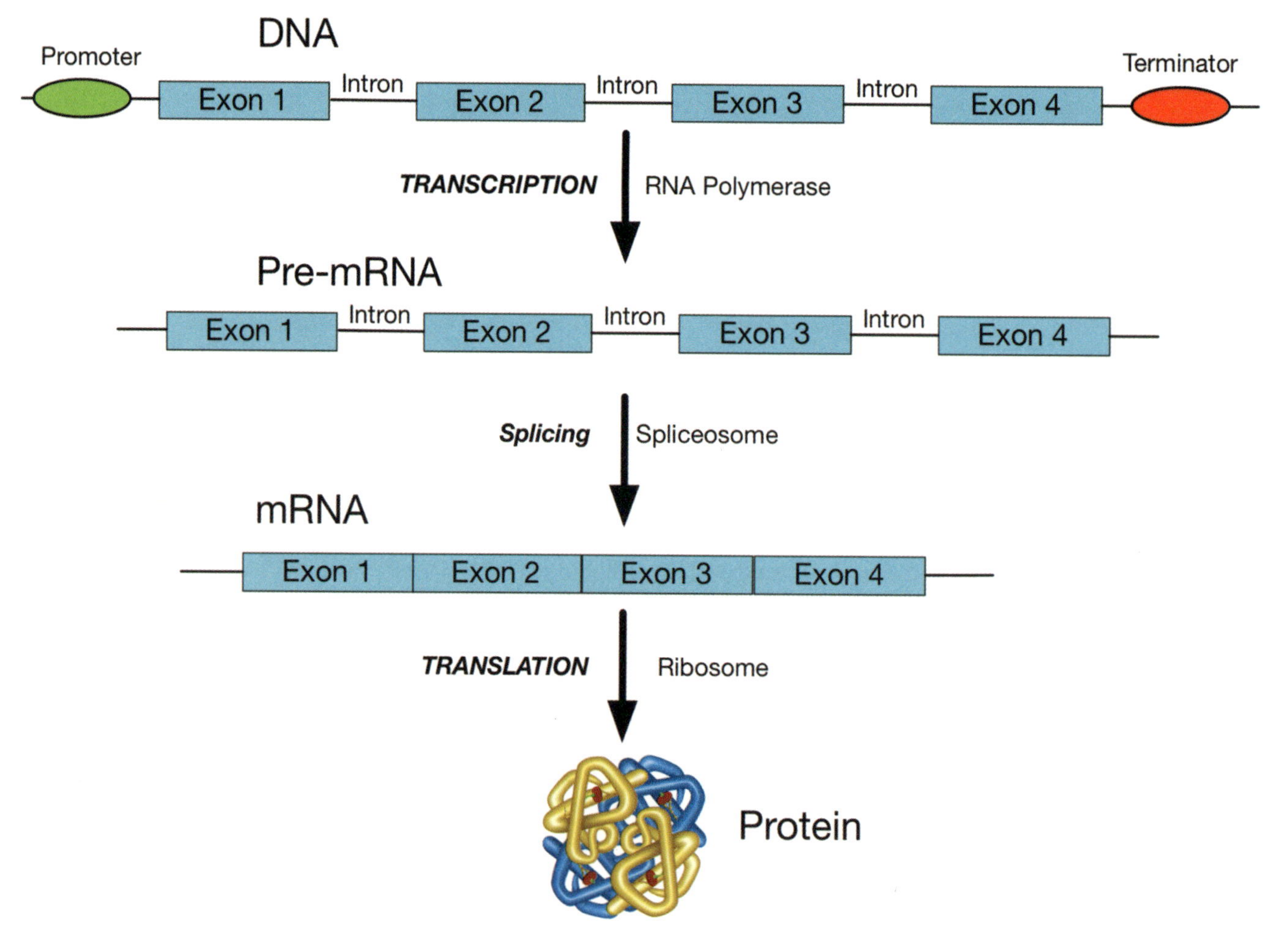

Fig. 3-9.

proteins, thereby increasing the number of kinds of proteins. For example, antibodies are formed by the combination of a variety of heavy and light protein chains. So, transposons, introns, and post-translational protein modification all contribute to the generation of millions of protein types. This is a great advantage in adding diversity to the genome, but may also contribute to the appearance of new disorders.

Spliceosomes are large molecular complexes, mainly in the nucleus, composed of a number of **small nuclear RNAs (snRNA)** and proteins, which together form **small nuclear ribonucleoprotein** complexes called **snRNPs** (affectionately referred to as "snurps"). A spliceosome removes ("snips off") introns from the transcribed pre-mRNA to form the processed mature mRNA molecule.

Small nuclear RNAs (snRNAs) are not the same as **microRNAs (miRNAs)**. While snRNAs (about 150 nucleotides long on average) are involved in removing introns from pre-mRNA, miRNAs (about 22 nucleotides long) bind to and silence corresponding mRNAs, thereby inhibiting translation and protein production and modifying gene expression. MiRNAs have been implicated in the generation of some cancers and cardiovascular disease. Some miRNAs originate in genes, while others originate in introns, transposons, or elsewhere.

Transcription Factors

The nuclear DNA in every cell of the body is the same. In order for the multitude of different kinds of cells in the body to function, it would be chaos to simply have all the DNA in each cell transcribe at once into the same RNA and the same proteins, making every cell alike. In order for cells and tissues throughout the body to display their unique differences, some genes have to be turned off and other genes on, in a timely manner, depending on the cell's specialization. What regulates this, and why doesn't all the DNA transcribe at once?

TRANSCRIPTION, TRANSLATION, PROTEIN FOLDING

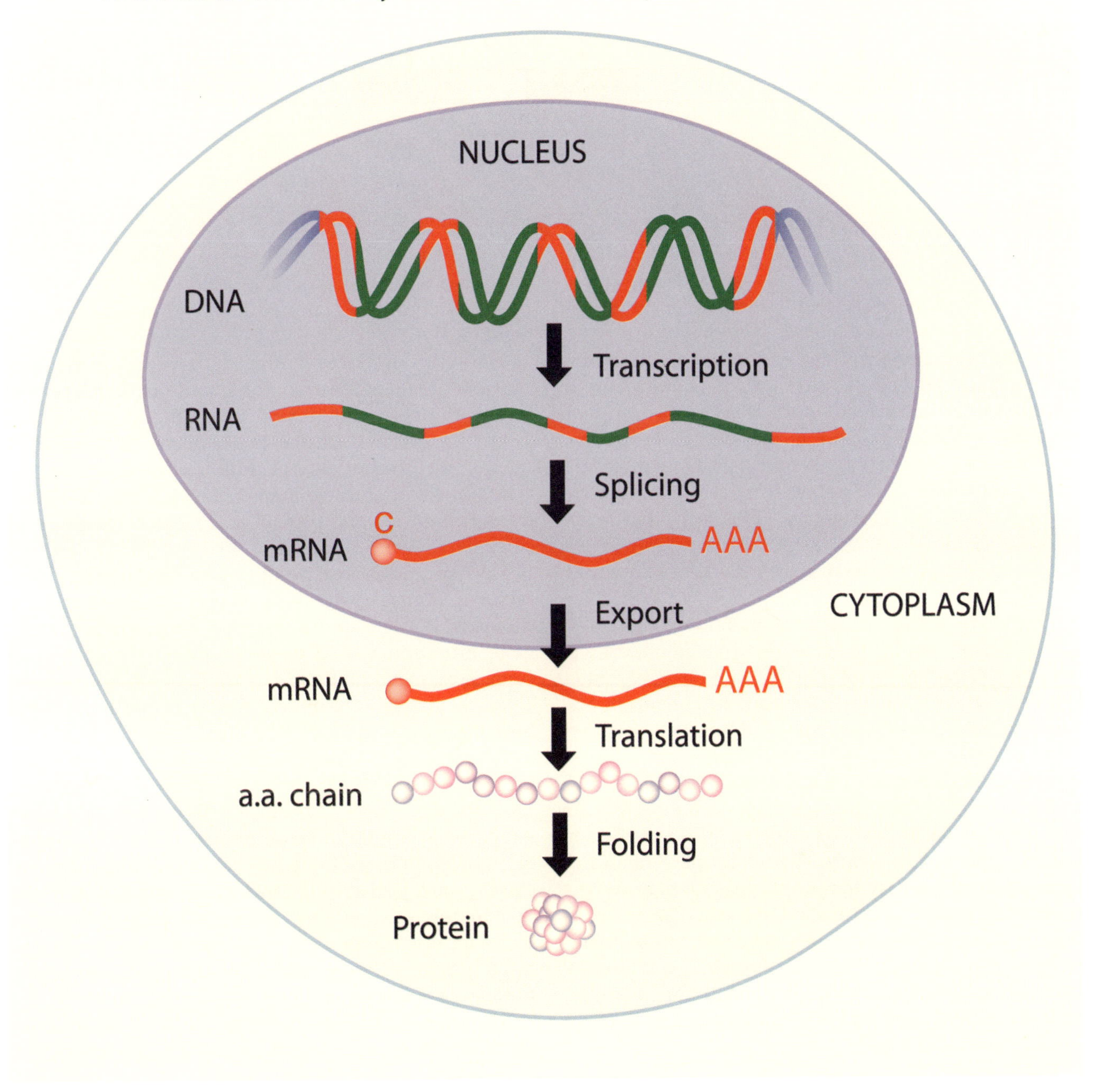

Fig. 3-10.

Transcription factors are proteins (called **activators** and **repressors**) that respectively increase or decrease the degree to which a particular gene or group of genes is expressed. There are thousands of different transcription factors. They enable cells with the same genome to differ in the amount and kind of protein they produce. In addition to binding to promoters on the DNA (where they facilitate the binding of RNA polymerase and help promote transcription), transcription factors also bind to **enhancers** or **silencers** (part of the DNA) on the chromosome to modify the degree of transcription. Activator transcription factor proteins bind to enhancer DNA and increase gene expression; repressor transcription factor proteins

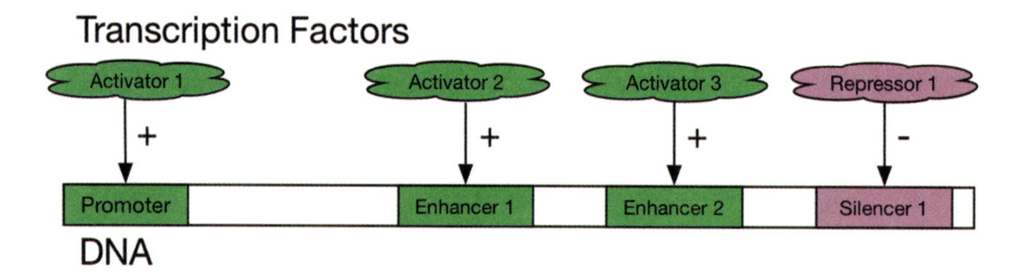

Fig. 3-11.

bind to silencer DNA and decrease transcription (**Figure 3-11**).

Figure 3-11. Transcription factors that are activators can promote gene expression (+) by connecting to promoters or enhancers, while repressors suppress gene expression (-) by binding to silencers. Promoters, enhancers, and silencers are all part of the DNA molecule, while activators and repressors are transcription factor proteins. (A mnemonic: "**TRA**nscription factor **DES**tination is DNA." I.e. **T**= **T**ranscription factor; **R** = **R**epressor; **A** = **A**ctivator; **D** = **D**NA; **E** = **E**nhancer; **S** = **S**ilencer.)

Promoters and terminators, as mentioned, lie at one end or the other of the gene, marking the boundaries of the gene, dictating where transcription begins or ends. While promoters initiate transcription, enhancers and silencers (there are hundreds of thousands of them, more than the number of genes) increase or decrease the degree of transcription, depending on which transcription factors in that particular cell interact with them. An enhancer can bind with several transcription factors. And a gene may have multiple enhancers. Conversely, one type of transcription factor may act on a number of genes. A combination of transcription factors may be needed to turn the gene on or off. An **insulator** is a segment of DNA generally found as a boundary between an enhancer and the promoter of an *untargeted* gene, helping to insure that the untargeted gene is not inadvertently expressed.

Even more fine-tuning of gene expression occurs through **coactivators** and **corepressors**, proteins (usually) that bind to activators and repressors respectively to further enhance or decrease their activity.

Unlike promoters, enhancers and silencers, while also made of DNA, do not need to be near the gene of interest or even on the same chromosome. They affect the gene by bending the shape of its DNA when they are bound by transcription factors and can affect the promoter region from a distance. DNA is not a rigid structure but is more like a wet noodle that can bend easily.

Transcription factor proteins bind to DNA and control the rate of transcription into RNA. Transcription factors can turn off the very genes that created them. This provides a feedback mechanism for controlling the quantity and quality of transcription factors in the cell. Transcription factors are often regulated by other transcription factors, providing a number of levels of control over gene expression.

DNA mutations can produce abnormal transcription factors. **Figure 3-12** lists some of the resultant diseases. While each disease may be due to a specific gene mutation that directly produces the abnormal transcription factor, there can also be an indirect cause, through a second mutated gene that influences the first one.

Gene mutations that result in defective transcription factors (**Figure 3-12**) may result in a number of diseases with facial dysmorphism, growth problems, intellectual disability, and other dysfunctions. Disorders of transcription factors have also been implicated in cancer, since transcription factors are implicated in the up and down regulation of genes, including those genes that suppress tumors, enhance cell division, or engage in DNA repair (unrepaired genes are more susceptible to cancerous mutation).

Mutations to an enhancer may also be responsible for certain diseases, e.g. *preaxial polydactyly, thalassemia, Van Buchem disease*, and *X-linked deafness*.

Mutations to a silencer may be implicated in *asthma, allergies*, and *fascioscapulohumeral muscular dystrophy*.

Mutations to coactivators may be implicated in Parkinson disease and Type II diabetes mellitus.

The structure of the DNA may remain normal, but the proper function of the cell's transcription factors may be altered by environmental influences. For instance, *thalidomide*, a drug notorious for producing limb malformations in the embryo, disrupts a wide network

FIGURE 3-12. SOME DISORDERS OF TRANSCRIPTION FACTORS
• Alzheimer disease • Amyotrophic lateral sclerosis • Aniridia • Autoimmunity • Campomelic dysplasia • Cancer (breast/prostate/colorectal/renal/lung/nasopharyngeal/melanoma/acute lymphoblastic leukemia) • Cardiovascular diseases • Charcot-Marie-Tooth disease • CHOPS syndrome • Congenital central hypoventilation syndrome • Congenital heart disease • Cornelia de Lange syndrome • Developmental verbal dyspraxia • Diabetes • Down syndrome with acute megakaryoblastic leukemia • Duane-radial ray syndrome • Inflammatory disease • Ivic syndrome • Li-Fraumeni syndrome • Lujan syndrome • Multiple sclerosis • Nail-patella syndrome • Neurologic disorders • Obesity • Opitz-Kaveggia syndrome • Osteoarthritis • Prostate cancer • Rett syndrome • Roberts syndrome • Rubinstein-Taybi syndrome • X-linked deafness • X-linked dyserythropoietic anemia and thrombocytopenia • X-linked thrombocytopenia • X-linked autoimmune-allergic dysregulation syndrome

of transcriptional factors, but does not appear to act by causing gene mutation. Regretfully, once the damage has been done, removing the drug cannot reverse the condition.

There are other vital parts of the chromosome, besides promoters/terminators and enhancers/silencers, that affect the degree of DNA transcription to RNA. **Histones** are proteins (not DNA) around which DNA wraps on the chromosome, thereby compacting the DNA (**Figure 3-13**). The combination of histone "spools" and wrap-around DNA is termed a **nucleosome**. **Chromatin** is the group packaging of nucleosomes. Generally, histones, when *demethylated* (i.e. methyl groups are removed from them) or *acetylated* (acetyl groups are added to them) can increase DNA transcription by "unwinding the spool." The unwinding allows transcription factors to bind to the DNA, thus increasing gene expression (**Figure 3-13**). Some transcription factors can catalyze the methylation/demethylation or acetylation/deacetylation of histones.

Figure 3-13. Histones. Unwinding of DNA around histones increases the accessibility of modifying transcription factors to DNA, and increases gene expression.

In general:

- *Methylation* of DNA inactivates gene expression; *demethylation* of DNA activates gene expression.
- *Acetylation, phosphorylation,* and *demethylation* of histones activate gene expression by unwinding the DNA that is wound around histones and allowing greater DNA accessibility. *Deacetylation* of histones rewinds DNA around the histones and inactivates gene expression (the DNA is rewrapped), as does *methylation* (although histone methylation in some circumstances can activate gene expression).

Activation and inactivation of gene expression in a given cell is important in that it enables one cell to distinguish its activities from another, so that all cells are not doing the same thing. The activation

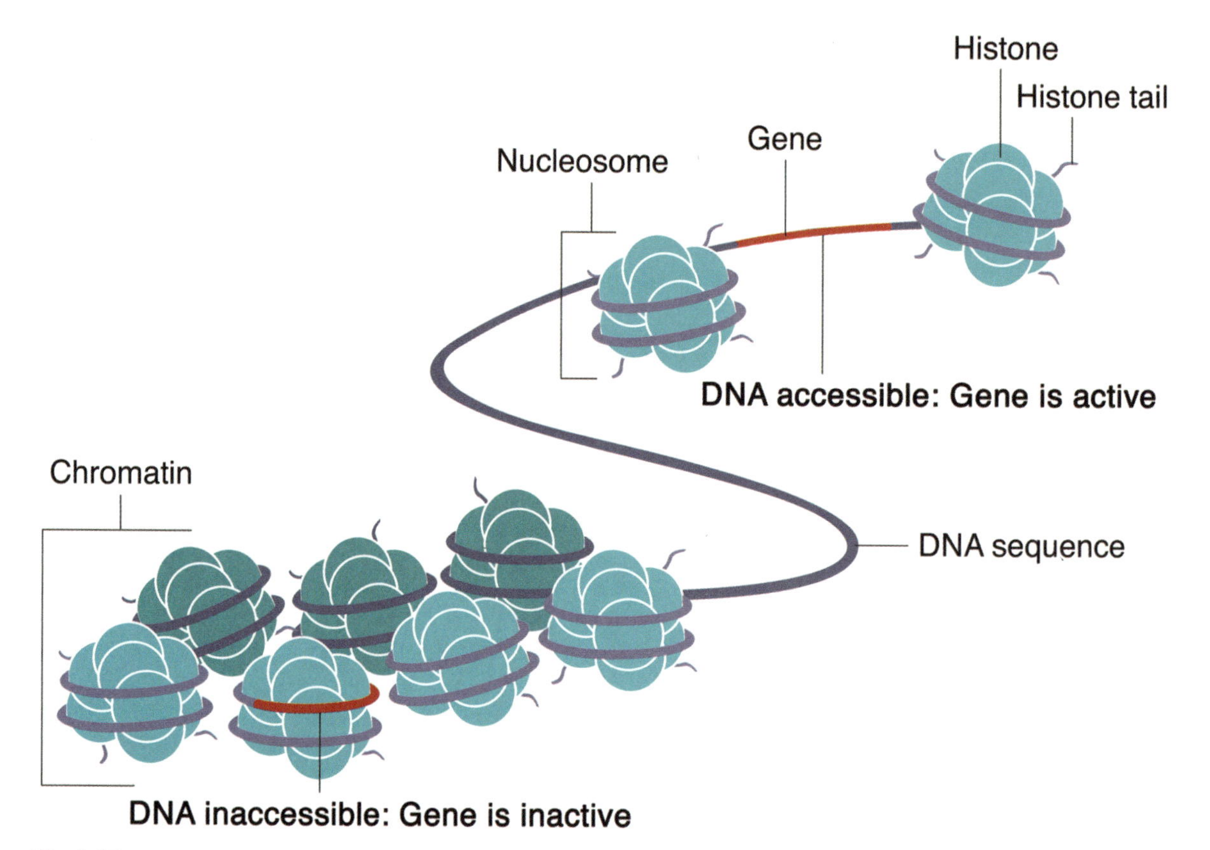

Fig. 3-13.

and deactivation can extend through mitosis, so that the state of one cell carries on its phenotype to its offspring (i.e. the factors altering gene expression are *bookmarked* during mitosis). It is now known that such bookmarking can carry over in germ cell Meiosis for certain traits acquired in the environment, even though this does not involve an actual change in the DNA sequence. Hence, the inheritance of acquired characteristics for certain traits (see **Chapter 5. Epigenetic Disorders**).

Figure 3-14 lists a number of diseases believed to be due to abnormally functioning chromatin with defective enzymes involved in the methylation/demethylation and acetylation/deacetylation of histones.

We then see a number of levels of control that can alter gene expression:

1. Transcription factors normally bind to promoters, enhancers, and silencers (all of which are composed of and lie within DNA) to alter gene expression without altering the DNA nucleotide sequence. Certain transcription factors can also catalyze the acetylation/deacetylation of histones. Mutant transcription factor proteins can result in defective transcription factor binding and a number of diseases.

FIGURE 3-14. CHROMATIN DISORDERS

- Alpha-thalassemia myelodysplasia syndrome
- Brachydactyly
- CHARGE syndrome
- Claes-Jensen XLMR syndrome
- Histone lysine demethylase deficiency
- Cockayne syndrome
- Coffin-Lowry syndrome
- Coffin-Siris syndrome
- Floating-Harbor syndrome
- Genitopatellar syndrome
- ICF syndrome
- Kabuki disease
- Kleefstra syndrome
- Nicoladis-Baraitser syndrome
- Retinal degeneration
- Rett syndrome
- Rubinstein-Taybi syndrome
- SBBYS (Say-Barber-Biesecker-Young-Simpson) syndrome
- Schimke immuno-osseous dysplasia
- Schinzel-Giedion syndrome
- Siderius XLMR syndrome
- Sotos-like syndrome
- Tatton-Brown Rahman syndrome
- Wolf-Hirschhorn syndrome

2. Histones normally block transcription factors from binding to DNA. Mutant enzymes that interfere with histone methylation/demethylation and acetylation/deacetylation are a cause of transcription factor malfunctioning.
3. Environmental (epigenetic) factors can affect methylation/demethylation and acetylation/deacetylation, thereby altering gene expression.
4. MicroRNA (miRNA), one of the products of DNA transcription, has an indirect silencing effect on DNA expression by binding to mRNA and cancelling out the expression of the mRNA.

Translation of RNA into Proteins

The DNA in a single chromosome may contain millions of nucleotide base pairs. Chromosomal DNA is divided into segments of nucleotides, called genes. Each gene is responsible for the eventual production of a particular protein. It used to be thought that each gene corresponded to a single enzyme, but an enzyme is just one kind of protein, which catalyzes (speeds up) specific chemical reactions without being changed. DNA, though, makes all the other kinds of proteins, which have completely different functions other than those of enzymes. (**Figure 3-15**).

Mutations in DNA can impair any of the protein functions, resulting in many kinds of diseases. Protein functions include:

- *Enzymes*. Enzymes are a large and enormously important category of proteins. All enzymes are proteins (except **ribozymes**, which are ribosomal enzymes made of RNA). There are thousands of enzymes, each of which catalyzes a different biochemical reaction. There are over 500 diseases that arise from enzyme defects, so-called **inborn errors of metabolism**, including those involving the metabolism of carbohydrates, lipids, steroids, amino acids, DNA, lipoproteins/glycolipids/glycoproteins, heme, and the generation of energy. **Figure 3-16** lists a representative sampling of hereditary enzyme disorders. Nearly all are autosomal recessive.
- *Motion*. Muscle contraction occurs when filaments containing the proteins *actin* and *myosin* slide along one another. In cells with *cilia* (short hair-like structures found in large numbers in a cell and which may be motile or nonmotile) or *flagella* (which are long, motile, and few in number on a cell, e.g. sperm), these proteins facilitate movement.

FIGURE 3-15. FUNCTIONS OF PROTEINS

- Enzymes
- Motion
- Structure
- Control of gene expression
- Growth substances
- Immune mechanisms
- Clotting mechanisms
- Hormones/Messengers
- Transport

Mutations in actin and myosin genes alter the structure of actin and myosin and are responsible for a number of *myopathies*, which are manifest with varying degrees of muscle weakness due to faulty muscle contractility. *Muscular dystrophy*, of which there are some 30 different forms, results from a defective gene for producing *dystrophin*, a protein that helps build and maintain muscle cells. Without normal dystrophin, the muscle cells eventually die.

Disorders of cilia (*ciliopathies*, of which there are at least 35, involving over 180 known ciliary protein genes), involve the disruption of cilia on a number of cell types, with conditions as diverse as *polycystic kidney disease*, *retinal degeneration* (e.g. *retinitis pigmentosa*), brain anomalies, and skeletal malformations (**Figure 3-17**). Cilia can sense mechanical stimuli such as fluid flow, move fluid along the cell surface (e.g. the respiratory tract epithelium, which clears mucus), and are also involved in chemosensation and photoreception. Flagella are involved in sperm motility, and their dysfunction, which can have a genetic origin, is one cause of male infertility.

- *Structure*. *Collagen* is a widespread protein that provides a structural framework of intercellular tissue support in connective tissue, cartilage, bone and other tissues. *Elastin* is a stretchable support protein for organs and tissues such as the skin, lungs, ligaments, blood vessels and heart. *Keratin* is the tough protein of fingernails and hair.

Hereditary forms of collagen disease include *Ehlers-Danlos syndrome*, *Marfan syndrome*, and *osteogenesis imperfecta*. All involve defects in the structure of collagen.

Hereditary diseases of elastin may result in loose sagging skin; aortic aneurysm; aortic stenosis and dissection; emphysema; and joint problems.

Hereditary diseases of keratin may manifest as fragile bleached hair; skin blisters; and thickening or hornlike growth of the skin on the hands and feet.

Protein defects may also occur in the structure of ribosomes. The ribosome, while still continuing to function, may only translate certain RNA messages. In some cases these protein defects have been associated with anemia and bone marrow failure in early life, and a later increased risk of cancer.

FIGURE 3-16. SOME INHERITED INBORN ERRORS OF METABOLISM DUE TO ENZYME DEFECTS AR = AUTOSOMAL RECESSIVE; AD = AUTOSOMAL DOMINANT; XR = X-LINKED RECESSIVE; XD = X-LINKED DOMINANT; GSD = GLYCOGEN STORAGE DISEASE
Carbohydrate Metabolism Aldolase deficiency (AR) Deficiency of branching enzyme (Type IV GSD; Anderson disease) (AR) Deficiency of debranching enzyme (Type III GSD; Cori Disease) (AR) Deficiency of lysosomal alphaglucosidase (Type II GSD; Pompe Disease) (AR) Diphosphoglyceromutase deficiency (AR) Enolase deficiency (AD) Essential fructosuria (fructokinase deficiency) (AR) Essential pentosuria (deficiency of xylitol dehydrogenase) (AR) Galactokinase deficiency (defect in galactose 1-P-uridyl transferase) (AR) Glucose-6-phosphate dehydrogenase deficiency (XR) Glucose-6-phosphatase deficiency (Type I GSD; Von Gierke's Disease) (XR) Glucose phosphate isomerase deficiency (AR) Hereditary fructose intolerance (deficiency in fructose 1-phosphate aldolase) (AR) Hexokinase deficiency (AR) Lactase deficiency (AR) Liver phosphorylase deficiency (Type VI GSD; Hers Disease) (AR) Liver phosphorylase kinase deficiency (Type VIII GSD) (AR) Muscle phosphorylase deficiency (Type V GSD; McArdle disease) (AR) Phosphofructokinase deficiency (Type VII glycogen storage disease) (AR) Phosphoglycerate kinase deficiency (AR) Pyruvate kinase deficiency (AR) Triosephosphate isomerase deficiency (AR)
Lipid Metabolism Carnitine deficiency (mitochondria) Niemann-Pick disease (AR) Refsum disease (AR) Respiratory distress syndrome (AR)
Steroid Metabolism Cholesterol desmolase deficiency (AR) 3-beta-hydroxysteroid dehydrogenase deficiency (AR) 21-hydroxylase deficiency (XR, XD) 11-beta hydroxylase deficiency (AR) 18-dehydrogenase and 18-hydroxylase deficiency (AR) 17-alpha hydroxylase deficiency (AR) 17,20 lyase deficiency (AR)
Amino Acid Metabolism Albinism (tyrosinase deficiency) (AR) Alkaptonuria (defect in homogentisate oxidase) (AR) Argininosuccinic aciduria (decreased argininosuccinase activity) (AR) Carcinoid tumor (AD) Cystathioninuria (deficiency of cystathioninase) (AR) Cystinuria (AR) Ehlers-Danlos Syndrome (AD, AR) Fanconi syndrome (AR) Formiminotransferase deficiency (AR) Hartnup disease (AR) Histidinemia (lack of histidase) (AR) Homocystinuria (defect in cystathionine synthase) (AR) Hypermethioninuria (decrease in methionine adenosyl transferase) (AR) Hypervalinemia (defect in valine transaminase) (AR) Isovaleric acidemia (AR) Maple Syrup Urine Disease (AR) Osteogenesis imperfecta (AD, AR, XR) Phenylketonuria (deficiency of phenylalanine hydroxylase) (AR) Tyrosinemia (AR)

DNA Metabolism
Adenosine deaminase deficiency (AR)
APRT (adenine p-ribosyl transferase) deficiency (AR)
Gout (multifactorial)
Lesch-Nyhan syndrome (deficiency of hypoxanthine P-ribosyl transferase) (XR)
Orotic aciduria (AR)
PRPP synthetase overactivity (XR)
Purine nucleoside phosphorylase deficiency (AR)
Pyrimidine 5'-nucleotidase deficiency (AR)
Xanthine oxidase deficiency (AR)

Energy Generation
Fumarase deficiency (AR)
Pyruvate dehydrogenase deficiency (AR, X-inactivation)

Lipoprotein, Glycolipid, Glycoprotein Metabolism
Abetalipoproteinemia (AR)
Fabry disease (XR)
Familial combined hyperlipidemia (AD)
Familial dysbetalipoproteinemia (AR, AD)
Familial hypercholesterolemia (AR)
Familial hypertriglyceridemia (AD)
Familial hypobetalipoproteinemia (AR)
Familial lipoprotein lipase deficiency (AR)
Gaucher disease (deficiency of beta-glucosyl ceramidase) (AR) and related diseases—Hunter syndrome (XR), Hurler disease (AR), Scheie disease (AR), I-cell disease (AR), Maroteaux-Laury disease (AR), Morquio syndrome (AR), Mucolipidosis VII disease (AR), multiple sulfatase deficiency (AR), and Sanfilippo A and B diseases (defects in enzymes that degrade mucopolysaccharides) (AR)
Krabbe disease (globoid leukodystrophy)(AR)
LCAT deficiency (AR)
Tangier disease (AR)
Tay-Sachs disease (deficiency in hexosaminidase A) (AR)

Heme Metabolism
Alpha thalassemia (AR)
Beta thalassemia (AR)
Coproporphyrinogen oxidase deficiency (variegate porphyria) (AD)
Crigler-Najjar syndrome (AR)
Dubin-Johnson syndrome (AR)
Ferrochelatase (heme synthetase) deficiency (AD, AR)
Gilbert syndrome (deficiency in UPD-glucuronyl transferase) (AD)
Gray baby syndrome (underdeveloped microsomal UDP-glucuronyl transferase) (AR)
Protoporphyrinogen oxidase deficiency (AD)
Rotor syndrome (AR)
Sickle cell anemia (AR)
Thalassemia major (Cooley's anemia) (AR, AD)
Thalassemia minor (AR)
Uroporphyrinogen decarboxylase deficiency (porphyria cutanea tarda) (AD, AR)
Uroporphyrinogen I deficiency (acute intermittent porphyria) (AR)
Uroporphyrinogen III cosynthase deficiency (congenital erythropoietic porphyria) (AR)

- *Control of gene expression.* We have seen that transcription factors (activators and repressors) and histones are proteins with important functions in controlling gene expression. Many diverse diseases, including cancer, autoimmunity, diabetes, cardiovascular diseases, neurological disorders, and obesity are connected with disorders of genes that produce transcription factor proteins (**Figure 3-12**). One transcription factor may disrupt a number of genes at once and affect a number of biological functions. Mutations involving histone structure have been implicated in a number of types of cancer.
- *Growth substances.* Certain proteins promote cell division and growth of tissue in the embryo and adult or are involved in programmed cell death (**apoptosis**). Mutations in these genes can hasten cell death, as in certain neurodegenerative diseases

FIGURE 3-17. EXAMPLES OF CILIOPATHIES

- Alstrom syndrome
- Asphyxiating thoracic dysplasia
- Bardet-Biedl syndrome
- Cranioectodermal dysplasia
- Ellis-van Creveld syndrome
- Joubert syndrome
- Kartagener syndrome
- Leber congenital amaurosis
- McKusick-Kaufman syndrome
- Meckel-Gruber syndrome
- Nephronophthisis
- Orofaciodigital syndrome
- Polycystic kidney disease
- Senior-Loken syndrome
- Short rib-polydactyly syndrome

(e.g. *Huntington disease*, *Alzheimer disease*, *Parkinson disease*, and *amyotrophic lateral sclerosis*), or cause abnormal growth, as found in cancer.

- *Immune mechanisms*. The immune system controls the spread of infections and eliminates foreign material. Genetic diseases of the immune system may affect the development and function of antibodies, B or T lymphocytes, phagocytic cells, complement, and cytokines or their receptors, all of which are vital to immune function. Examples are *asthma*, *ataxia telangiectasia*, *autoimmune polyglandular syndrome*, *Burkitt lymphoma*, *diabetes Type I*, *DiGeorge syndrome*, *familial Mediterranean fever*, and *immunodeficiency with hyper-IgM*.

Autoimmune diseases, a kind of hyperactivity of the immune system, where the body attacks itself as if it were a foreign substance, may sometimes run in families, involving multiple genes and environmental factors. Sometimes the inheritance is only a *tendency* to show autoimmunity, which may appear differently in different people in the same family, e.g. *rheumatoid arthritis*, *multiple sclerosis*, *lupus*, and *thyroid disorders*.

- *Clotting mechanisms*. The clotting factors, (e.g. fibrinogen and thrombin) are proteins. Inherited diseases of clotting include the *hemophilias* and *von Willebrand disease*. The specific disease depends on the particular clotting cascade factor that is defective or missing.
- *Hormones/Messengers*. Many of the hormones are peptides or proteins (e.g. insulin, growth hormone, adrenocorticotrophic hormone). Genetic diseases can involve deficiency or excess of those hormones.
- *Transport*. Proteins may act as carriers to transport other molecules. For instance, hemoglobin is a protein that carries O2 in red blood cells; myoglobin transports O2 in muscle cells; transferrin transports iron; thyroglobulin-binding protein transports thyroxin. Albumin forms the largest proportion of plasma protein. Albumin carries various hormones, iron, heme, vitamins, bilirubin, free fatty acids, Ca^{++}, rare metals and many drugs, enabling them to be soluble in plasma. Attachments to albumin render many molecules inactive. Albumin also has an osmotic effect that helps to maintain the blood volume.

FIGURE 3-18. SOME DISORDERS OF MEMBRANE TRANSPORT PROTEINS

- Cystic fibrosis
- Cystinuria
- Glucose/galactose malabsorption
- Hereditary spherocytosis
- Hereditary elliptocytosis (ovalocytosis)
- Hereditary stomatocytosis
- Hyperkalemic periodic paralysis
- Liddle syndrome
- Myotonia congenita
- Nephrogenic diabetes insipidus
- Niemann-Pick disease
- Wilson disease

There are also *membrane transport proteins*, which transport substances, such as ions and small and large molecules, including proteins, across biological membranes, as well as membrane structural proteins. Genetic mutations may result in deficiencies of these proteins (**Figure 3-18**).

- *Storage*. *Ferritin* is a protein that stores iron in the liver. *Hemochromatosis*, of which there are several genetic forms (usually recessive), is marked by the excess accumulation of iron in the body, with symptoms that may include fatigue, abdominal pain, loss of sex drive, heart flutter, memory fog, and loss of hair.

Messenger, Transfer, and Ribosomal RNA

Messenger RNA (mRNA) carries the genetic code as a sequence of base (adenine, uracil, guanine, or cytosine) triplets, called **codons** (**Figure 3-19**). Each codon triplet corresponds to a specific amino acid. An **anticodon** is the triplet site on a transfer RNA molecule that recognizes the messenger RNA codon, enabling the transfer RNA, with its attached amino acid, to fit properly on the messenger RNA. (It is not the amino acid that attaches to the messenger RNA codon, but the anticodon section of the tRNA molecule that attaches to the mRNA codon.) (**Figures 3-19** and **3-20**).

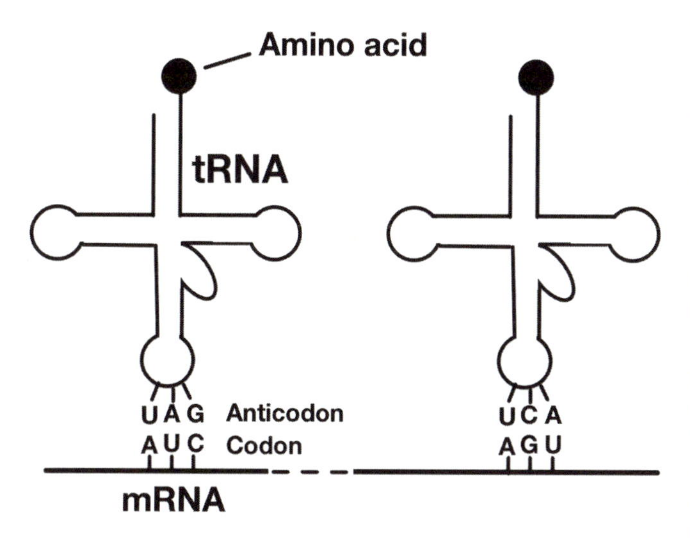

Fig. 3-19.

Figure 3-19. Alignment of tRNA anticodons with mRNA codons.

Figure 3-20 shows how messenger RNA (mRNA), transfer RNA (tRNA), and ribosomal RNA (rRNA) work together in the translation of mRNA into proteins.

A given codon on the mRNA molecule matches only one specific amino acid. It has to be that way; otherwise a messenger RNA could no longer code for only one specific protein. However, several different codons can code for the same amino acid. There are 61 codons on mRNA that match only 20 amino acids.

Certain unique codon triplets on messenger RNA function as **START** (AUG) or **STOP** (UAG, UAA, or UGA) signals, signifying where the protein construction should start or stop.

A **sense codon** is one that actually codes for an amino acid. A **nonsense codon** (a STOP codon) does not code for an amino acid and causes the translation to terminate, leaving a shorter, incomplete protein.

Relatively rarely, RNA, after RNA transcription and prior to translation, in itself may be changed (**RNA editing**) and not have the same sequence generated by the DNA. This can occur through the deletion, insertion, or substitution of nucleotides in RNA. Such editing modifies gene expression and contributes to genetic diversity, but has been implicated in the development of certain neuromuscular, lung, and liver diseases, and cancer.

Protein Modification

Different proteins can be formed by modifying the proteins that come off the ribosome assembly chain, whether by trimming off parts of them or adding new parts. For instance, the inactive protein *trypsinogen* is trimmed to form the active digestive enzyme *trypsin*. *Insulin* is formed by cleavage of a larger,

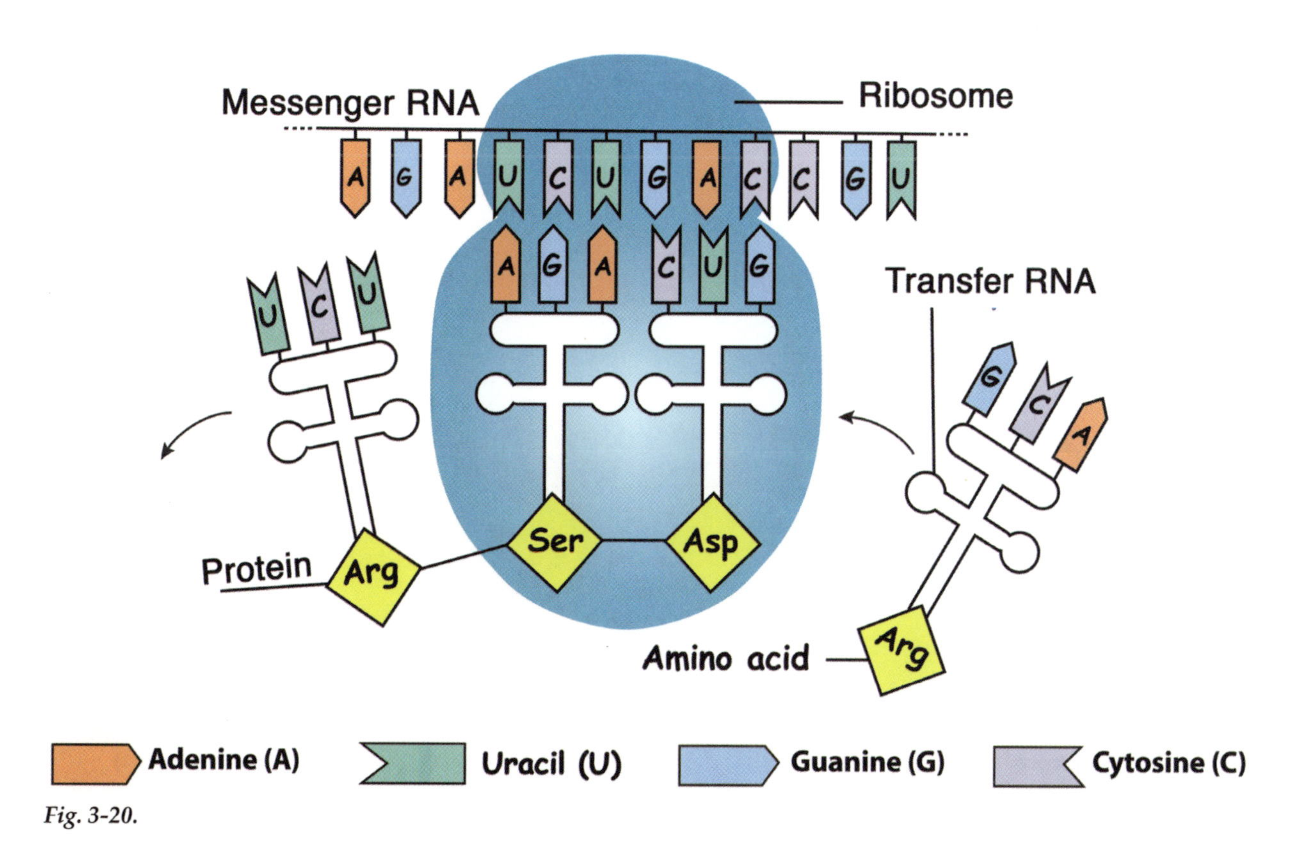

Fig. 3-20.

precursor *proinsulin* molecule. The hydroxyproline and hydroxylysine proteins in collagen are formed by adding hydroxyl groups to proline and lysine on their precursor protein molecules. Sometime carbohydrate side chains are added to the protein to form a *glycoprotein*. Lipids can also be added (*lipoproteins*), as well as other groups (e.g. iron in hemoglobin).

An important aspect of a protein that makes it work is its shape. The particular amino acid sequence in a protein affects the shape of the protein. So do environmental influences, such as pH, temperature, reducing/oxidizing agents, and other small molecules. The protein shape, apart from its chemical structure, affects how it will fit and connect with cell surface receptors or other molecules. Small changes in the shape can affect the degree of the protein's effectiveness, especially if it is an enzyme; small changes in enzyme structure can increase or decrease the rate of a biochemical reaction. The **primary structure** of a protein is simply the sequence in which its amino acids are strung out (**Figure 3-21**). The **secondary structure** of a protein is its shape when it is laid out folded as sheets, spirals or loops. The **tertiary structure** of a protein is a further modification of 3D structure where parts of the protein connect with other parts through hydrogen bonding, disulphide and other kinds of bonds. The **quaternary structure** of a protein, if it has one, is the shape it forms in conjunction with other polypeptide chains. For example, collagen is composed of intertwining strands of collagen molecules; hemoglobin's quaternary structure is made of four interlocked polypeptide subunits. **Chaperone proteins** are a special type of protein that assists with proper protein folding both during and after translation.

Figure 3-21. Primary, secondary, tertiary, and quaternary structures of proteins.

Faulty protein folding has been implicated in a number of diseases, including:

- Alpha-1 antitrypsin deficiency
- Alzheimer disease
- Amyloidosis
- Creutzfeldt-Jakob disease
- Cystic fibrosis
- Gaucher disease
- Huntington disease
- Multiple system atrophy
- Parkinson disease
- Sickle cell disease
- Other neurodegenerative diseases

Mitochondrial Inheritance

Inheritance of parental characteristics is not just from nuclear DNA genes. There are separate **mitochondrial genes** with their own unique DNA (which is a double helix, but not paired as are the nuclear alleles, and is arranged as a continuous circle). Since each egg cell contains many mitochondria, but sperm cells contain none (or a few that are not used), inheritance of mitochondrial DNA comes totally from the mother.

Mitochondrial DNA exists as multiple copies, not only 2-10 copies of mitochondrial DNA in each mitochondrion, but many more, since each cell may have 100-1000 mitochondria. There is only one (diploid) set of nuclear DNA in each cell. however. Mitochondria are considered haploid, since their DNA is not paired.

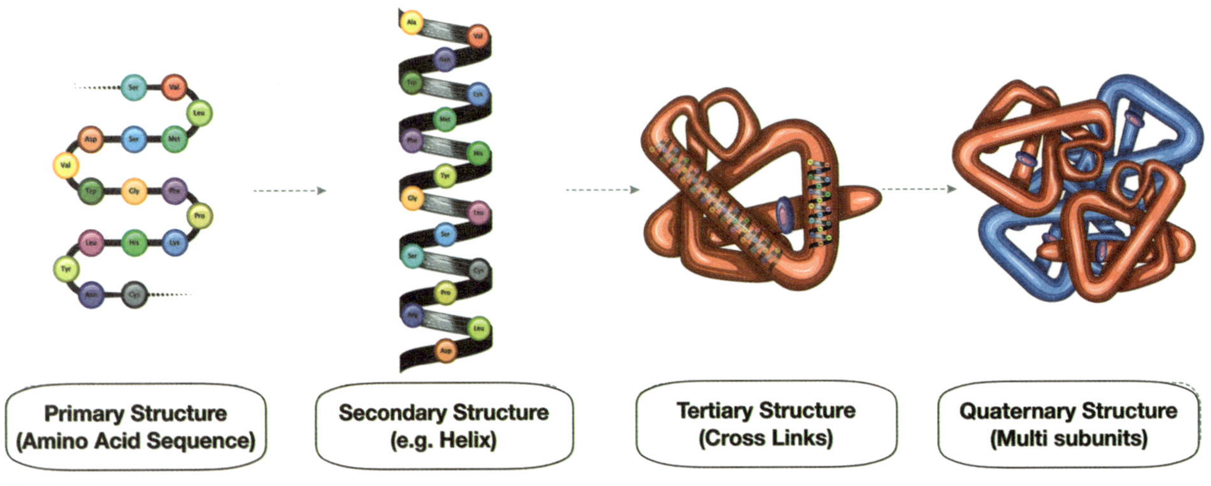

Fig. 3-21.

Mitochondria produce ATP for the energy needs of the cell. Mitochondrial DNA (**mitDNA**) mutations thus commonly affect brain and muscles, which have a high need for ATP, but disease manifestation may be diverse (**Figure 3-22**).

Carnitine, an amino acid synthesized from lysine and methionine, plays an important role in mitochondria function and energy balance by transporting fatty acids across the inner mitochondrial membrane. It also increases the utilization of carbohydrates. Disorders that involve carnitine deficiency, e.g. inadequate intake, enzyme deficiencies in the carnitine metabolic pathway, or excess losses through diarrhea, dialysis, or diuresis, are commonly treated with carnitine supplements.

One in 5,000 people has a *primary* mitochondrial DNA disease, which is always maternally inherited. Most people with mitochondrial disease have an inherited form, but sometimes the mutations in mitochondrial DNA may occur spontaneously. Mitochondrial dysfunction may also be *secondary*, resulting from mutations in DNA genes that produce proteins that affect mitochondrial function. It may be clinically difficult to determine if a mitochondrial disease is primary or secondary.

Mitochondrial mutations frequently have variable degrees of expression since there are thousands of mitochondria in cells and only some may be affected. A number of factors can cause variable expressivity, including not only genetic but environmental and lifestyle factors as well.

Genetic problems are thus not always simply due to a nuclear DNA mutation. They can result from many other factors that participate in the regulation of DNA replication, transcription of DNA into RNA, editing of RNA, translation of RNA into proteins, and post-translational protein modification, as well as mutations in mitochondrial DNA.

FIGURE 3-22. DISORDERS OF MITOCHONDRIA

Mitochondrial Disorders	**Symptoms of Mitochondrial Disorders**
Primary mitochondrial disorders	Autism-type behaviors
• Diabetes mellitus and deafness	Cardiac, liver, or kidney diseases
• Leber's hereditary optic neuropathy	Dementia
• Leigh syndrome	Diabetes
• MELAS syndrome	GI disorders
• Mitochondrial DNA depletion syndrome	Growth delay
• Mitochondrial myopathy	Infection
• Myoclonic epilepsy with ragged red fibers	Lactic acidosis
• Myoneurogenic gastrointestinal encephalopathy	Mental retardation
• Neuropathy, ataxia, retinitis pigmentosa, and ptosis	Muscle weakness
	Neurologic problems
Secondary mitochondrial disorders	Respiratory problems
• Alzheimer disease	Thyroid dysfunction
• Cancer	Vision/hearing problems
• Cardiovascular disease	
• Chronic fatigue syndrome	
• Diabetes	
• Huntington disease	
• Lou Gehrig disease	
• Muscular dystrophy	
• Reye syndrome	
• Sarcopenia	

Part II. When Things Go Wrong

With all the complexity during DNA replication and transcription into RNA, RNA translation into proteins, and protein modification, there are numerous areas where things can go wrong, and they do, often with different treatments needed to fit the particular problem.

A genetic mutation is a change in DNA nucleotide sequence or chromosome structure. It can be single point mutation, which affects a single nucleotide base pair, or a broader mutation that spans a number of nucleotides or whole genes. There can be even larger changes that affect the gross structure or number of chromosomes. Or there can be multifactorial disorders, involving multiple mutations and environmental factors that influence gene expression.

A gene can have hundreds of different mutations, resulting in qualitative or quantitative differences in the phenotypic abnormalities that arise. *Qualitatively*, different mutations of the same gene can cause totally different phenotypes. *Quantitatively*, different mutations of the same gene can result in variations of the same phenotype. For instance, the same gene that gives rise to sickle cell anemia gives rise to beta-thalassemia with a different mutation to the same gene. There are some 1700 different mutations of the single recessive cystic fibrosis (CFTR) gene, some of which do not cause disease while others do in different degrees, depending on the particular mutation. A mutation can affect the quantity of normally functioning protein, or affect the function of the protein itself. Environmental factors may also play a role in the severity of the disease.

Sometimes, there can be mutations to a different gene that result in the same phenotypic abnormality. For example, the same connective tissue disease, *Ehlers-Danlos*, can arise from mutations in different genes on different chromosomes; and there are autosomal-dominant, autosomal-recessive, and X-linked inherited forms of the disease. One reason for multiple causes of the same disease is that there may be a biochemical chain reaction which, when the particular enzyme at any step is interrupted, leads to the same deficit at the end. Each step may use a different enzyme, produced by a different gene (Figure II-1)

Figure II-1. In a multistep pathway, where each enzyme arises from a different gene, a defect in any one of the genes may result in the same deficiency of the product, or backup accumulation of the substrate on which the enzyme acts, with similar clinical manifestations.

In some cases, the exact mutation can result in different phenotypic effects in different people, depending on the modifying influence of other mutations elsewhere in the genome or the influence of environmental factors. The same gene mutation may not affect the child to the same degree as the parent.

In Meiosis, sometimes one gamete may receive two

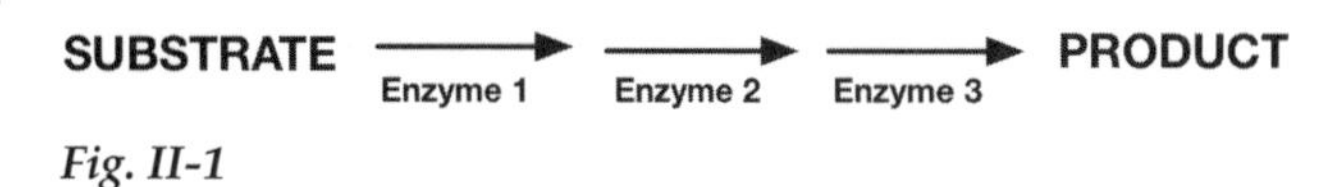

Fig. II-1

chromosomes and the other receives none. This can lead to a condition where the person has too many or too few chromosomes. A person with two X chromosomes (XX) is female, and a person with one X and one Y chromosome is male. However, a person may be XXY (*Klinefelter syndrome* in males, having an extra X chromosome), or X0 (*Turner syndrome* in females, having one less X chromosome). The Y chromosome determines male sex.

There are much rarer conditions where an entire extra set of chromosomes is present so that every cell contains 69 or 92 chromosomes (**triploidy** or **tetraploidy** respectively), rather than 46 (**Figure 4-1**).

Apart from such gross abnormalities in the number of chromosomes, which are easily seen microscopically when the chromosomes are laid out into a visible **karyotype**, there are other changes that are smaller and may be harder to see. Sometimes, even though the chromosome number is normal there may be:

a. A deletion of part of a chromosome
b. A duplication of part of a chromosome
c. An inversion of part of a chromosome
d. Translocation of a segment from another chromosome that is not part of the pair.

With special staining techniques, one can microscopically see a number of such anomalies. Special staining highlights characteristic visible band patterns on each chromosome that can be used to identify changes to a chromosome (apart from other identifying markers, such as the size of the chromosome and the position of its centromere).

There are finer abnormalities at the molecular level that are even harder to see microscopically. These include:

a. Deletion or duplication of one particular gene or part of a gene, such as a single nucleotide in the DNA sequence
b. Substitution of one nucleotide base for another

Such microscopic anomalies require more refined laboratory techniques to discover. Understanding the underlying cause of the patient's problem can help guide therapy.

Epigenetics is a branch of genetics concerned with the environmental factors that influence phenotype while not altering the genotype. Epigenetic influences surprisingly may sometimes be inherited. Some people require inheritance as part of the *definition* of epigenetics. Others more broadly include in the definition of epigenetics any (non-mutational) changes that affect gene expression throughout life (maintained through multiple mitotic divisions), regardless of whether they are inherited. Epigenetic influences include external environmental influences such as diet, obesity, drugs, tobacco and alcohol intake, pollutants, stress, and sleep irregularities, as well as internal bodily influences, such as hormones, cytokines (small proteins involved in cell communication), and growth and cell differentiation factors. **Epigenomics** (*genome* refers to the overall DNA that a person has) is the study of the **epigenome**, the overall network of chemicals and other factors around DNA that affect its expression without mutating the DNA itself. The epigenome may affect DNA methylation/acetylation, and the functions of histones and microRNA (microRNA silences RNA functioning), which in turn change DNA expression rather than change the actual DNA code sequence. Epigenetic factors can alter the functioning of transcription factors.

4

Chromosomal Abnormalities

Too Many or Too Few Chromosomes

Aneuploidy is having extra or missing chromosomes. In **triploidy** (**Figure 4-1**) there is an entire extra set of chromosomes. In a human that would be a person with 69 chromosomes, generally not compatible with life. Triploidy results in miscarriage or dying within a few days of birth, except in cases where there is a somatic mutation and only some cells in the body contain the extra set (a **mosaic**). In plants, **polyploidy** (e.g. triploidy, quadriploidy, and higher) is often an advantage, leading to an enhancement in the fruit or vegetable, such as increased vigor or larger size.

Figure 4-1. Examples of polyploidy.

Trisomy refers to one extra chromosome. Thus, there may be three of chromosome 21, rather than two, so the person has 47, rather than 46, chromosomes.

In humans, only trisomy 13, 18, and 21 are compatible with life. Trisomy 21 is the most common, found in *Down syndrome* (**Figure 4-2**). *Patau syndrome* (trisomy 13; **Figure 4-3**) and *Edwards syndrome* (trisomy 18) are less common and accompanied by severe physical and intellectual defects. There are probably many more examples of trisomy in humans that go undetected because they are damaging enough to cause a miscarriage. It is estimated that about 20% or more of pregnancies end in a miscarriage, which seems very high, but many miscarriages likely occur very early in pregnancy before a woman knows that she is pregnant.

Figure 4-2. Down syndrome. The most common form has an extra chromosome 21.

Figure 4-3. Patau syndrome. There is an extra chromosome 13.

Down syndrome is significantly more common with pregnancies in older women. Part of the reason may be that the cell division of Meiosis II is not completed until later in life, during fertilization; the cell is older. There are also hereditary genetic diseases, though, that are correlated with advanced paternal age (> age 40) (**Figure 4-4**).

A chromosome abnormality can present as a whole extra chromosome or extra DNA content in a chromosome, where the chromosome number is the same but one of the chromosomes is longer than usual, containing a duplication of part of the chromosome (**Figure 4-5D**). Alternatively, an entire chromosome may be missing (the person has 45 rather than 46 chromosomes), or a part of a chromosome may be missing (a **deletion**) (**Figure 4-5C**). Duplications tend to be less harmful than deletions, partly because an extra chromosome segment may simply supply more of a needed protein, while a deletion may eliminate a needed protein, with a resulting disease. Also, **monosomy** (only one chromosome of a pair exists) may unmask a recessive lethal trait that would not be manifest if there were a corresponding normal allele. It is believed that of the 20,000 or more genes, everyone carries several genes that are pathological but are not manifest because they are recessive.

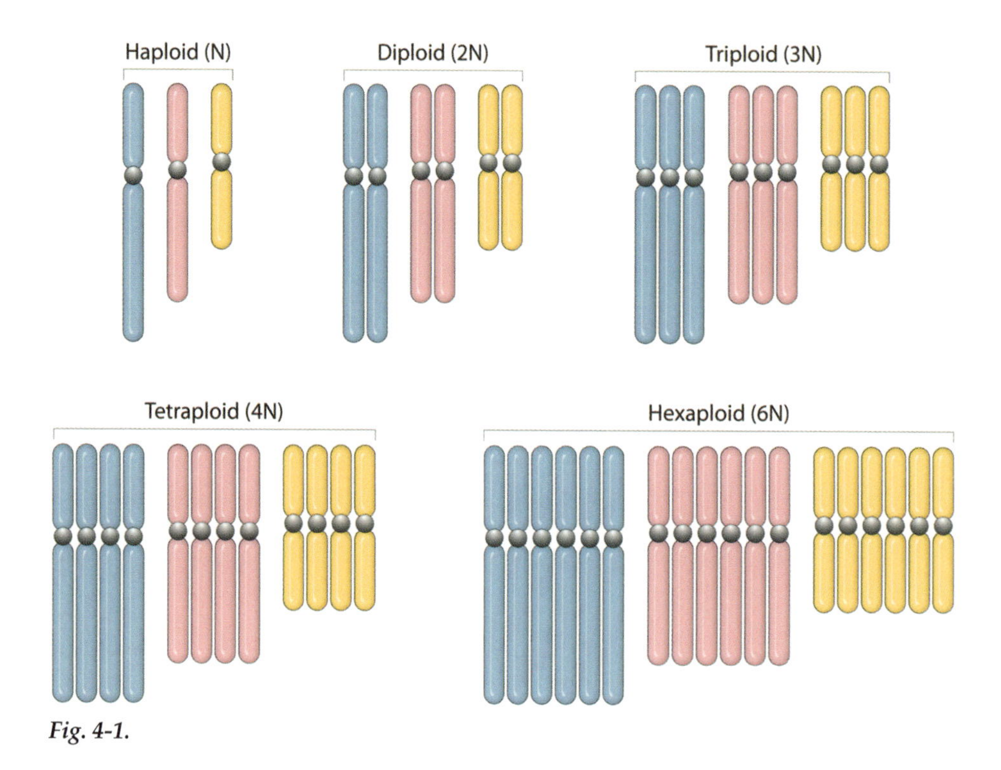

Fig. 4-1.

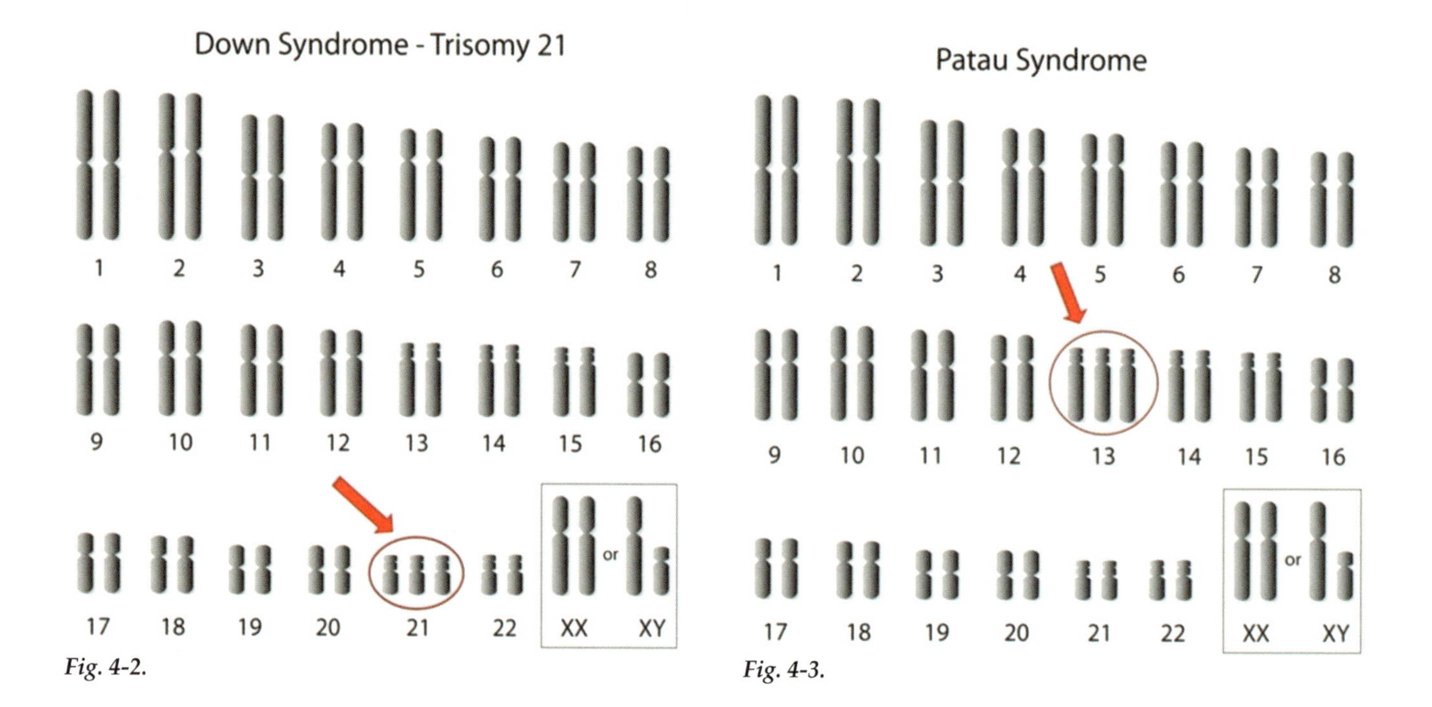

Fig. 4-2.

Fig. 4-3.

In the past it was thought that genes were almost always present as two copies, paired alleles on a pair of chromosomes. We now know that the gene copy number can be increased, and the copy segments can be long, encompassing whole genes, or shorter. Extra gene copies can arise de novo or be inherited. They are often normal variants and may even have a beneficial effect. It is thought, for instance, that having extra copies of the CCL3LI gene can protect against contracting HIV and AIDS. However, gene copies are often susceptible to mutation. Extra copies have been implicated in susceptibility to a number of cancers and mental, ocular, and neurologic disorders, including Alzheimer disease and schizophrenia.

FIGURE 4-4. DISEASES CORRELATED WITH INCREASED PATERNAL AGE
• Achondroplasia • Apert syndrome • Childhood acute lymphoblastic leukemia • Down syndrome • Leukemia • Marfan syndrome • Multiple endocrine neoplasia types 2 and 3 • Myositis ossificans • Neurofibromatosis • Osteogenesis imperfecta • Pfeiffer syndrome • Schizophrenia • Thanatophoric dysplasia

Figure 4-5. Chromosome inversion, translocation, deletion, and duplication.

Translocation of Chromosome Segments

A translocation can occur within the same chromosome pair, where two parts of the chromosome exchange with one another, or the exchange can occur between two nonrelated (**non-homologous**) chromosomes (called **reciprocal translocation**) (**Figure 4-5B**).

A translocation can be balanced or unbalanced. In a **balanced translocation**, sections of two chromosomes have switched places, but there is still the correct number of genes. In an **unbalanced translocation**, there is a resulting duplication or deletion of genes. A balanced translocation commonly has a normal phenotype, but the phenotype may be pathological if the break was in the middle of a gene, thereby impairing the gene's function. *Chronic myelogenous leukemia* (CML), for instance, is the result of a balanced translocation between chromosomes 9 and 22. The resulting *Philadelphia chromosome* 22 is unusually short (**Figure 4-6**), while chromosome 9 is longer. A fused gene in CML produces an overactive tyrosine kinase. Tyrosine kinase is a cell enzyme that promotes cell proliferation, cell survival and migration, and when overexpressed, as occurs in chronic myelogenous leukemia in the Philadelphia chromosome, is associated with the development, progression, and recurrence of the cancer. Treatment includes administration of tyrosine kinase inhibitors.

Figure 4-6. The Philadelphia chromosome. An abnormal fused gene (red band) on chromosome 22 produces a new protein that promotes excessive production of white blood cells in *chronic myelogenous leukemia*.

A **Robertsonian translocation** occurs when the translocation involves two chromosomes whose centromeres are very close to the end of the chromosome (called **acrocentric**). In that case, a single long chromosome results, along with a very small chromosome, which often disappears and is clinically inconsequential. The total chromosome number is then 45, but essentially there is a full, balanced complement of genes, and the person is normal. The person's progeny may not be, however, if they inherit the long abnormal chromosome from the affected parent in addition to a normal one from the other parent, which would be equivalent to having an extra chromosome (**Figure 4-7**), an unbalanced state. Robertsonian translocations can

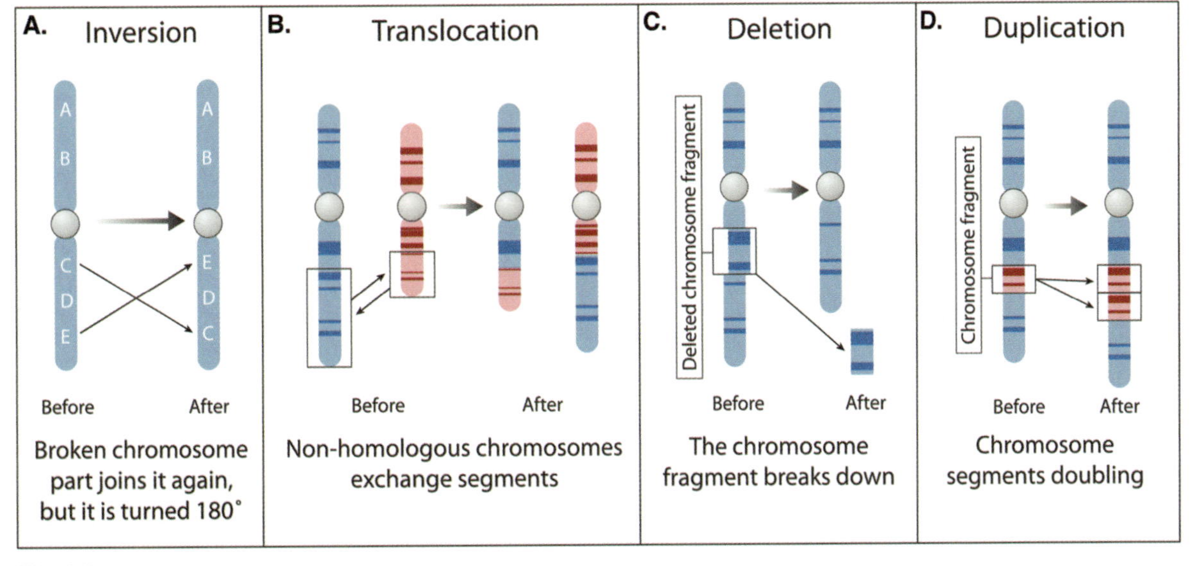

Fig. 4-5.

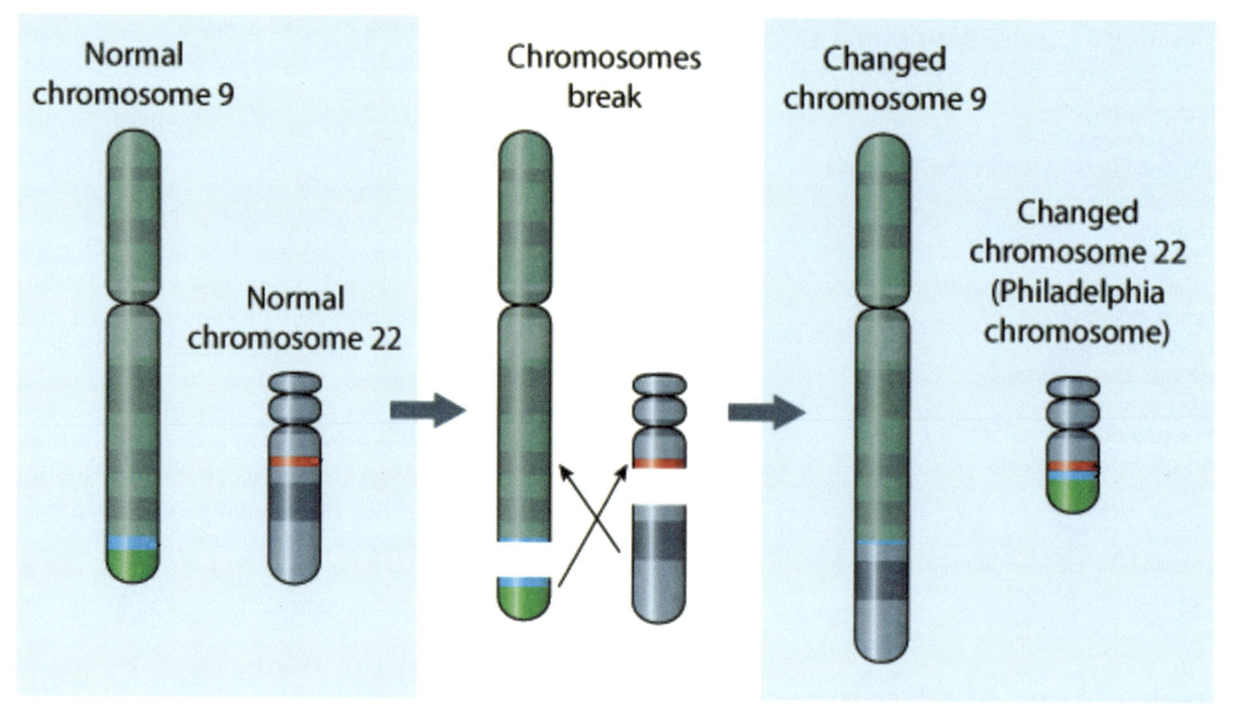

Fig. 4-6.

involve chromosomes 13, 14, 15, 21, and 22, which are acrocentric.

Figure 4-7. Robertsonian translocation.

A child with Down syndrome may have the usual spontaneous trisomy 21 (an extra chromosome) or a Robertsonian variant. Even though the parent may be phenotypically normal, it is important to test the parent of a child with Down syndrome for a Robertsonian translocation, which could be transmitted to other children they may plan to have. About 5-6% of children with Down syndrome have a Robertsonian translocation trisomy 21.

Although there is little evidence of a man with Down syndrome fathering a child (perhaps due to defective sperm or lack of sexual interaction), about 30-50% of women with trisomy Down syndrome are believed fertile. Although it is rare that they reproduce, they can become pregnant with about a 35-50% risk of having a child with Down syndrome.

A gene's expression may differ depending on what area the gene is next to (**position effect**). For example, its new location may be next to a heterochromatic region (condensed regions of the chromosome where gene activity is repressed), which inhibits the expression of nearby genes.

Translocations may be implicated in conditions as different as cancer, mental retardation, and infertility.

Inversion of Chromosome Segments

Inversions of a part of a chromosome (**Figure 4-5A**), despite resulting in no change in chromosome number or number of nucleotides, may still be harmful if the break occurs in the middle of an essential gene, thereby rendering the gene functionless. *Inversion on chromosome 9* is most common. It may have no clinical significance, but in some cases has been associated with growth retardation, miscarriage, congenital anomalies, and cancer. Inversions on the X chromosome may result in *hemophilia*, *Hunter syndrome*, and certain cancers.

An inversion may or may not include the centromere. A **pericentric inversion** includes the centromere; a **paracentric inversion** does not (**Figure 4-8**). Most people with paracentric or pericentric inversions are healthy unless the inversion interrupts an important functioning gene. When the abnormal and matching normal chromosome undergo Meiosis I, however, and attempt to cross over (**synapsis**) to exchange chromosomal genes, the mismatched chromosome segments cross over unequally. This results in defective gametes, which cause genetic problems, including miscarriage, congenital anomalies, infertility, growth retardation, and cancer.

Figure 4-8. Paracentric and pericentric inversions.

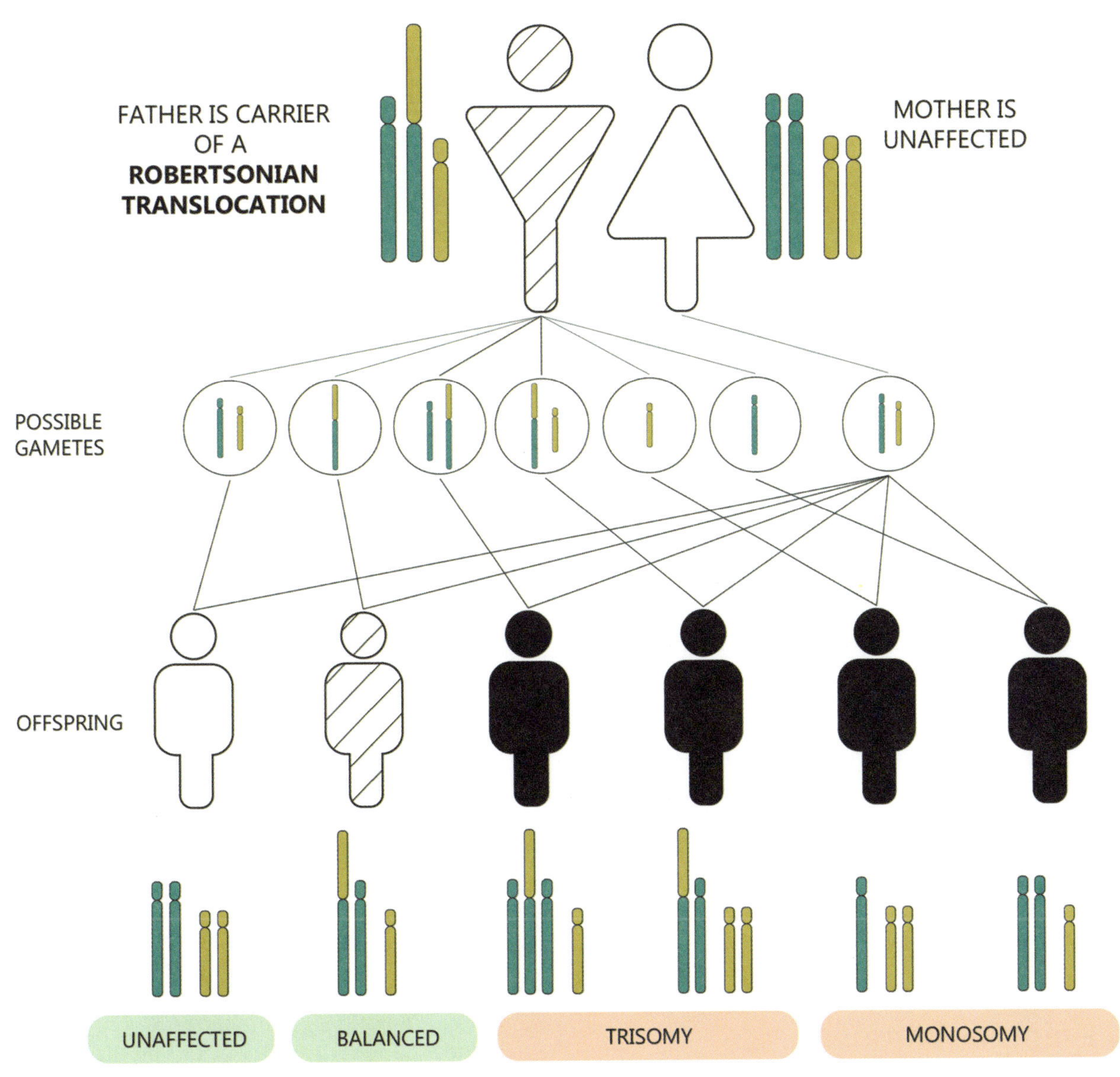

Fig. 4-7.

Duplications and Deletions of Chromosome Segments

While the number of chromosomes may be normal, a chromosome may have a segment that is duplicated or deleted (**Figure 4-5**). For example, in *cri du chat syndrome* there is a deletion of the short arm of chromosome 5, with resultant microcephaly, mental retardation, hypotonia, and a high-pitched catlike cry. It is easy to see how a deletion may result in a phenotypic defect, but a duplication may also be harmful. Too much of anything is not necessarily good. Segment duplications have been implicated in diseases as diverse as *autism*, *Crohn disease*, cancer, and a number of neurologic disorders.

Microdeletions and **microduplications** are chromosome changes too small to be detected by conventional light microscopy, but usually encompass more than one gene. They may be associated with developmental delays, learning disabilities, autism, epilepsy, and tumors, among other things.

PARACENTRIC AND PERICENTRIC INVERSION

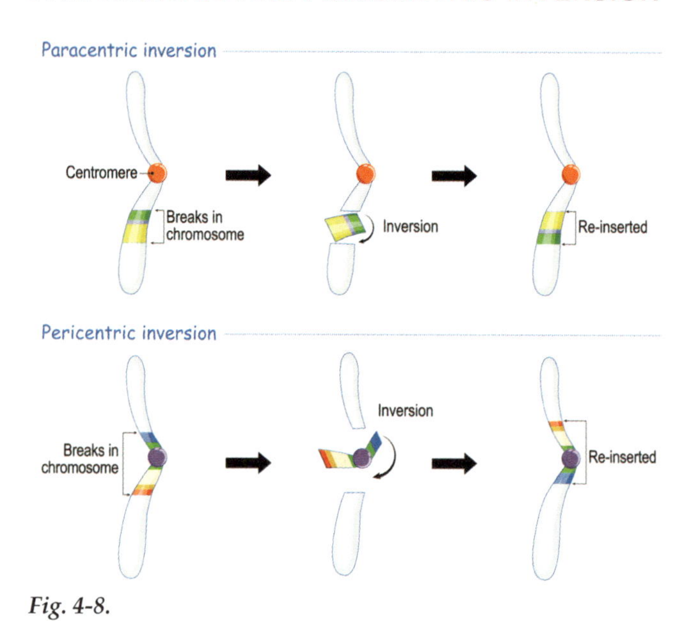

Fig. 4-8.

FIGURE 4-9. SOME DISEASES OF MICRODELETIONS AND MICRODUPLICATIONS	
Microdeletions	**Microduplications**
Angelman syndrome DeGeorge syndrome Miller-Dieker syndrome Neurofibromatosis Types I and II Prader-Willi syndrome Rubinstein-Taybi syndrome Smith-Magenis syndrome Williams-Beuren syndrome Wolf-Hirschhorn syndrome	Charcot-Marie-Tooth disease type 1A Microduplication 17p13.3 syndrome Microduplication 22q11.2 syndrome Potocki-Lupski (17p11.2) syndrome

DNA Point Mutations

A point mutation may present in various ways:

- A point mutation may involve the *deletion* of one base pair, a *duplication* of the base pair, or the *substitution* of one base pair for another. In some cases, this small change may be inconsequential. In other cases, the point mutation may be highly significant, as in sickle cell anemia, where a single wrong base pair can result in protein folding that is so abnormal that a significant disease results.
- The mutation may occur at any of a number of points in the same gene. Any one of the mutations may suffice to disable the resultant protein. For example, there are over 1700 known mutations to the cystic fibrosis gene. Although some of them have no effect, others result in the disease. The resulting phenotype may depend on the exact location in the gene and the influence of other genes or environmental factors.
- In some cases, the same phenotypic abnormality that results from a mutation to one gene may also result from a mutation to a different gene, even one that lies far away in a different chromosome pair. For example, there may be a complex chemical process in a cell with different steps that require a different enzyme at each step in the chain to function. A mutation that disables even one of the enzymes may be enough to disrupt the end product. Thus, a particular disease may result from a mutation to any of the genes responsible for one of the enzymes, however far away one gene may be from the other. For example, in *Ehlers-Danlos syndrome* (a connective tissue disease that results, among other things, in hyperelasticity of the skin), there are a number of known types, resulting from different genes on different chromosomes. Chemical reactions in a cell are like a vast spider web of contacts and intercommunications. Interrupting the web at different points may result in similar diseases, with variations.

A point mutation may have minimal if any effect if:

a. The same amino acid is coded for despite the DNA base substitution, as different triplets can sometimes code for the same amino acid (61 triplets code for 20 amino acids), resulting in the same protein (a *silent mutation*).
b. In a *neutral mutation*, the change occurs in a non-coding part of the gene (e.g. in an intron - **Figure 3-9**), or there may be a change in gene function, but not enough to cause a problem.
c. There may be a significant problem, but one that does not occur until later life.
d. Many mutations are harmful only in the homozygous state and will not be manifest as a phenotypic change in the heterozygous state.

In a **nonsense mutation**, a point change in a triplet mistakenly encodes for a STOP codon in a messenger RNA, resulting in a defective protein, since the translation of RNA stops at the STOP codon. For instance, one form of *familial adenomatous polyposis*, an inherited (autosomal dominant or recessive) disorder with colorectal cancer, has been found associated with a premature STOP codon in RNA (UAG, UAA, or UGA). Colorectal cancer in general may be due to a single mutation or to multiple gene mutations and contributing environmental factors.

In a **missense mutation**, a change in a nucleotide results in a different triplet that codes for a different amino acid, which will decrease the function of the resultant protein (**Figure 4-10**). The disease manifestation may be less than would occur with a deletion of the entire gene.

A **conditional mutation** affects the protein's function only under certain circumstances, e.g. a change in temperature, different metabolic state, light, or other mutations in the genome.

A **gain-of-function mutation** is one where a mutation (usually presenting as a dominant allele) results in a different protein with a new enhanced function over the normal protein. This contrasts with the more common **loss-of-function mutation**. The point mutation responsible for *sickle cell anemia*, for instance, can offer a gain by protecting against malaria. Other useful gain-of-function mutations include HIV resistance, lactose tolerance, and, of course, Spiderman. Gain-of-function mutations, however, can add harm too; more is not necessarily better. For instance, a gain-of-function mutation may result in a cancer by excessively stimulating the cell to divide. Mutations to the PCSK9 gene, which normally helps maintain the amount of blood cholesterol, may be implicated in *familial hypercholesterolemia*.

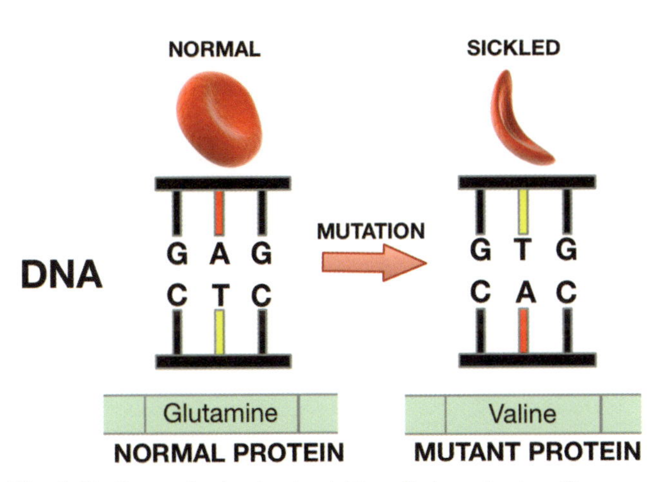

Fig. 4-10. Base substitution in sickle cell. A = adenine; C = cytosine; G = guanine; T = thymine.

A microdeletion or microduplication of a *single* base pair in an otherwise functioning gene can be especially harmful since it can throw the triplet codon sequence totally out of sync. This results in a **frameshift mutation**, where the whole sequence of amino acids is way off base (pun intended), with no resulting functioning protein. As an analogy, consider the following:

No mutation: THE CAT ATE THE RAT

Substitution of B for C: THE BAT ATE THE RAT (Altered sentence but not that way off)

Deletion of C: THE ATA TET HER AT (Frameshift, unintelligible)

Insertion of B after C: THE CBA TAT ETH ERA T (Frameshift, unintelligible)

5

Epigenetic Disorders

Epigenetics is the study of changes in gene expression that do not involve changes to the sequence of nucleotides in DNA. Sometimes the changes, surprisingly, may be inherited.

Whether or not you call something an *epigenetic disease* depends on your definition of *epigenetic*:

1. If you have a narrow definition of epigenetic that is restricted to environmentally acquired changes in gene expression that are *inherited* but do not result from changes in the DNA sequences, there are relatively few recognized diseases. There are those of **genomic imprinting**, which depend on whether the condition originates in the father or the mother, but are not sex-linked on the X or Y chromosome, e.g. *Angelman and Prader-Willi syndromes* (**Figure 5-1**), but there are a few that are not in that category:
   - In *Dutch hunger winter syndrome*, it was found that children who were born during the Dutch famine of 1944-45 were smaller than those born the year before the famine, and the discrepancy in size lasted for two generations, despite the return of food availability to normal.
   - Methylation patterns (see pages 19 and 23) in gene promoters have been correlated with a familial pattern of cancer.
2. If you expand the definition of *epigenetic* to also include non-inherited changes that originate after birth, without DNA mutation, and continue throughout life through all the cell's mitotic divisions, this greatly expands the list of epigenetic diseases.
3. If you broaden your definition still more to include environmentally acquired changes in gene expression in a cell that are not necessarily inherited or passed on through mitosis, but involve changes in the machinery that influence gene expression without DNA mutations, (e.g. abnormalities of histones, transcription factors, DNA methylation, and microRNA) this expands the list even more.
4. If you expand your definition even further to include *DNA mutations* that affect the machinery of gene expression, particularly transcription factors, this broadens the list still further, to the point of not really being useful as a definition of epigenetics.

This discussion will take the broader view of epigenetics (definition #2) that includes all factors, hereditary or not, that influence the expression of the genes through successive mitoses without altering the DNA sequence. In this view, epigenetic changes can persist throughout a cell's lifetime, as part of a cell's normal function. Such epigenetic changes enable each cell to continuously express its own individuality and produce proteins specific to each cell.

The bottom line, for medical purposes, involves treatment approaches to either prevent the harmful environmental effects on gene expression, or administer drugs to reverse the abnormal gene expression.

Epigenetic mechanisms can influence gene expression at all levels of cell activity, including transcription, translation, and protein modification. A disease that has an environmental epigenetic cause has the potential to be reversed more easily than one that results from gene mutation.

Cellular metabolism is a juggling act that maintains a correct balance of activity for the particular cell (termed **homeostasis**). Examples where homeostasis is needed include acid-base balance, temperature, glucose metabolism, calcium levels, fluid volume, immune system activity, blood pressure, blood clotting, and cell growth and repair.

Many environmental epigenetic factors can interfere with this balance, including diet, obesity, tobacco, alcohol, physical activity, stress, poor sleep patterns, environmental pollutants, temperature, hypoxia, inflammation, infection, and radiation. All these can affect the expression of genes and homeostasis. A number of them can also result in gene mutations, e.g. certain chemicals, x-rays, UV light, food preservatives, pesticides, herbicides, and drugs.

Trace metals (e.g. zinc, iodine, fluoride, selenium, copper) and vitamins are epigenetic environmental factors in that they are needed not only for the functioning of many enzymes and other proteins, but also play a role in DNA methylation and histone modification.

DNA methylation and histone deacetylation generally result in gene repression (see pages 19 and 23), while DNA demethylation and histone acetylation result in gene activation. Micro RNAs (miRNA) alter gene expression by targeting and silencing complimentary mRNAs. All these epigenetic mechanisms suggest potential new avenues for prevention or treatment of cancer, cardiovascular disorders, teratogenic effects and other diseases that affect gene expression, by manipulating the epigenetic factor rather than the gene itself.

Regarding cancer, an epigenetic change that silences a tumor suppressor gene, such as a gene that normally prevents excessive cell division, or an epigenetic change that directly increases cell division could lead to the uncontrolled cellular division in cancer. When an epigenetic change turns off genes that help repair damaged DNA, this may secondarily lead to an increased accumulation of DNA damage and mutations that increase cancer risk. Therapy that focuses on these mechanisms may lead to breakthroughs in cancer treatment and are currently under active investigation.

Sometimes a disease may result from a gene mutation that is impossible (presently) to correct, but the way in which the gene is expressed holds the potential for an epigenetic treatment. For instance, *fragile X syndrome* (an X-linked dominant condition), while the result of gene mutation, acts through abnormally condensed histones, which hinder DNA expression. There is a potential to approach the problem clinically through **chromatin remodeling** (e.g. via demethylation/acetylation), which changes chromatin to a looser, less compacted state accessible to transcription factors. Chromatin remodeling in principle may be easier than altering the structure of the gene itself. Epigenetic therapy thus holds the potential for treating a number of common conditions, including cancer, diabetes, heart disease, and mental illness.

Not every hereditary disease has a single gene at fault. The expression of the disease may be multifactorial, involving several sets of genes and environmental influences. The combination of genetics and environment (nature and nurture) determine our phenotype. Examples of disease with multifactorial causes include

- Alzheimer disease
- Arthritis
- Asthma
- Autism
- Cancer
- Coronary heart disease
- Diabetes
- High blood pressure
- Obesity

Normally, the attachments to DNA of epigenetic factors are wiped clean in the embryo through a process called **reprogramming**. This "resets" the genes in the early embryo so that they are no longer under the restrictive influence of epigenetic factors. Hence, the multipotentiality of early embryonic cells. In order to be inherited, epigenetic factors must avoid reprogramming.

In mammals, about 1% of genes escape epigenetic reprogramming through a process called **genomic imprinting**. Genomic imprinting is a type of epigenetic inheritance that depends on the sex of the parent. For instance, in *Prader-Willi syndrome* (obsessive eating and obesity, delayed motor development, cognitive

FIGURE 5-1. DISEASES ASSOCIATED WITH GENOMIC IMPRINTING
• Albright hereditary osteodystrophy
• Angelman syndrome
• Beckwith-Wiedemann syndrome
• DIRAS3 breast and ovarian cancer
• Familial nonchromaffin paraganglioma
• Male infertility
• McCune-Albright syndrome
• Prader-Willi syndrome
• Pseudohypoparathyroidism
• Silver-Russell syndrome
• Transient neonatal diabetes mellitus

impairment), which *can occur in males or females*, there is a deletion in *paternal* chromosome 15. In *Angelman syndrome* (unusually happy excitable personality, developmental delays, intellectual disability), a different phenotype from Prader-Willi syndrome, there is the same deletion, but in the *maternal* chromosome 15. This disorder, too, can occur in males or females. The reason for calling these disorders epigenetic is that the key difference between the two conditions lies mysteriously in whether or not the parent is male or female. Both conditions are autosomal dominant. In these disorders the sex of origin of the chromosome affects the expression of the gene.

Unlike Prader-Willi and Angelman syndrome, which do not depend on the sex of the child who has the condition, but rather on whether the chromosome inheritance is from the father or mother, there are many diseases in medicine that affect the sexes unequally, even when both male and female harbor the same mutation. Autoimmune diseases (e.g. *lupus erythematosus, Sjogren syndrome, scleroderma), Alzheimer disease, osteoporosis, histrionic personality disorder, fibromyalgia, irritable bowel syndrome, idiopathic hypersomnia*, and obviously, gynecological disorders, are more common in women. More common in men are *autism, schizophrenia, antisocial personality disorder, substance use disorder, X-linked recessive disorders* (e.g. *color blindness, hemophilia A and B*), *abdominal aortic aneurysms*, and, obviously, diseases of the male reproductive system (e.g. *prostate cancer*). In these cases, the mutation may be autosomal, X-linked or Y-linked, or may be due to lifestyle differences, or hormonal difference between the sexes to account for the differences in disease frequency and presentation.

Uniparental disomy is a rare condition in which the 2 chromosomes in a pair come from the same parent, rather than 1 copy from each. *Angelman syndrome* (AS) and *Prader-Willi syndrome* (PWS), in addition to their usual cause involving gene deletion within chromosome 15, can also be caused by uniparental disomy, depending on which parent contributes the 2 chromosomes. If they come from the father, for instance, it is equivalent to the deletion in the mother's chromosome 15, and the child exhibits Angelman syndrome. Prader-Willi occurs when both copies of chromosome 15, even when they are normal, are inherited from the mother.

In genetics, transmission does not mean the same thing as expression. **Transmission** is the transfer of genetic information from one generation to another. **Expression** is how and whether those genes function in the production of the phenotype. A nonexpressed gene can still be transmitted. The expression of the same gene may change depending on environmental influences.

It is not always easy to prove that it is an epigenetic influence that is being inherited rather than a DNA mutation. To do so, there are several indicators that a condition is epigenetic:

- Show that there is no DNA mutation responsible for the condition, which may be quite challenging. An epigenetic condition, unlike a gene mutation, may more easily reverse itself with proper environmental changes. If it does reverse, this reversal may suggest an epigenetic cause.
- Show that the condition extends into the 4th generation and is not due simply to a temporary environmental change. For instance, in the Dutch hunger winter syndrome, the smaller size of the succeeding generations of children persisted even after the famine was over.

Epigenetic factors may help explain differences between monozygotic twins. The DNA may be the same, but epigenetic differences acquired after conception can alter phenotype. Identical twins are a useful research tool in learning to distinguish genetic vs epigenetic changes.

Lyonization is an interesting example of epigenetics. In Lyonization (named after geneticist Mary Lyon), there is a normal random inactivation in somatic cells (cells other than the reproductive cells) of one of the two X chromosomes of the female embryo during embryonic development. The inactivated X chromosome remains inactivated for subsequent divisions (but this is reversible in germ cells),

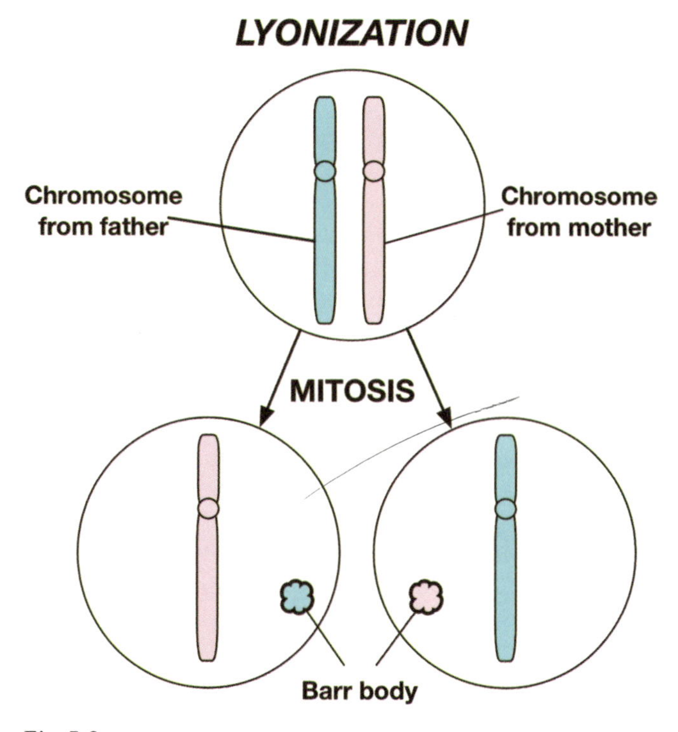

Fig. 5-2.

appearing as a crumpled up nonactive **Barr body** at the nuclear periphery (**Figure 5-2**). In some somatic cells, the Barr body originates from the father's X chromosome and in others from the mother's. Cells with a Barr body express their phenotype from the normal chromosome, and the body is thus a mosaic, with some X chromosomes in the female originating from the mother and others from the father. Each cell expresses only the X genes of the functioning X chromosome. A common example is the calico cat, in which the fur coloring is partly orange and partly white, in patches. Nearly all calico cats are female. They have a gene for the orange versus non-orange coloring on the X chromosome, which expresses itself in patches in only some cells in the mosaic. As a clinical example of X inactivation, someone with Klinefelter syndrome (XXY genotype) develops as a male because one of the X chromosomes forms an inactive Barr body, leaving the patient essentially XY (male).

Figure 5-2. Lyonization. In females, only one X chromosome is active in somatic cells due to inactivation of one of the chromosomes to form a Barr body. In some cells, the Barr body originates from the father; in other cells, from the mother. Only the active chromosome is expressed. It is an epigenetic phenomenon because it involves a difference in gene function without a change in DNA, depending on whether the active chromosome originates from the father or the mother, even if the genes are otherwise the same.

One may well ask, if one X chromosome is inactivated in the female, why doesn't the functioning chromosome with a "recessive" mutation manifest the disease? Actually, the heterozygous cells with the recessive mutation often do show the disease to some degree in females, but in half of the cells the functioning X chromosome comes from the mother and in other cells it comes from the father. Thus, the normal chromosome, which is active in half the cells, is able to help compensate for the abnormal gene.

Epigenetics can alter the number and degree of phenotypic defects in the population. A point mutation can have effects that differ among persons, based on different (epigenetic) environmental influences, as well as the influence of one or more genes on the expression of another. For example, people with a genetic predisposition to diabetes can be affected in different degrees depending on epigenetic factors as diet, exercise, and drugs.

It can be puzzling to try to understand how one mutation can affect so many diverse bodily functions. For example, in CHARGE syndrome, there may be retinal defects, heart problems, growth retardation, genital and ear abnormalities, and blockage of the back of the nasal passage. In part, diversity of symptoms and signs in a genetic disease may be due to the same elementary materials being used in different body areas to construct different things (e.g. collagen used as a building block in many diverse structures), just as a particular type of brick can be used in creating different kinds of buildings. Nature conservatively converts and reuses a lot of its elements. For instance:

- The complex porphyrin structures of chlorophyll and heme (the oxygen-carrying portion of hemoglobin) differ mainly in the inclusion of an iron molecule in heme and a magnesium ion in chlorophyll. Yet despite the close similarities in structure, the two molecules have very different functions.
- The molecule acetyl CoA, used to provide energy in the biological Krebs cycle, is also used as a building block to construct cholesterol, the bile salts, sex hormones, and prostaglandins.
- Glucose may be used to provide energy, or, if it loses a carbon atom, become a structural component of DNA and RNA, or it can combine with proteins and lipids to form glycoproteins and glycolipids, with their unique functions.
- Adenosine is not only part of the energy-providing molecule ATP (adenosine triphosphate) but also is incorporated into the structure of DNA and RNA.
- The sex hormones, cholesterol, chlorophyll, vitamins A, K, and E, and rubber all have a similar substructure consisting of the 5-carbon groupings referred to as *isoprenes*.
- The amino acids, when necessary, can be converted into carbohydrates or fats.

Also, a particular gene may be expressed in only certain tissues. For example, the TBX5 gene is selectively expressed in the heart and upper limbs. In *Holt-Oram syndrome* a mutant TBX5 gene results in cardiac malformation (septum defects) and absent wrist/thumb bones, with underdevelopment of the upper arm. There may be no other bodily defects because the gene is not expressed elsewhere.

Diverse organ systems, then, may make use of the same ingredients or alterations of them. Hence, the different signs and symptoms that may appear in a single disease, and the similarity of a number of signs and symptoms in different diseases. In addition, many of the metabolic pathways have housekeeping functions, found in many cells throughout the body. A defect in one reaction may affect the functioning of diverse cells, leading to diverse clinical conditions.

Also, there are genes that control the expression of a number of diverse genes in whole body areas, in a sense like an orchestra conductor controls the sections of the orchestra. Mutations to these genes thus can significantly affect many aspects of body structure and function at once.

6

Cancer

An estimated 38.4% of men and women (~ 1 in 3) will be diagnosed with cancer sometime in their lives. About 13% of women (~1 in 8) will be diagnosed with breast cancer. By age 80, about 80% of men have prostate cancer (but only 1 in 39 will die of it). About 4.5% of men and women (1 in 23) will develop colorectal cancer.

Gene mutations that result in cancer generally are those that promote excess cell division or those that inactivate tumor suppressors. Other mutations damage DNA repair genes or enable the tumor cell to avoid immune recognition and destruction.

Factors that can act as carcinogens include diet, smoking, some viruses and bacteria, radiation, some hormones, and environmental toxins. Some of these cause gene mutations and some are epigenetic, altering cell function without causing a change in the gene sequence.

Only about 5-10% of cancers are hereditary (**Figure 6-1**). Nonhereditary cancer, though, is still a genetic disease, for two main reasons:

1. The cancer genes arise from a local somatic DNA mutation later in life and therefore are not hereditary.
2. The genes may remain normal, but outside epigenetic factors may alter the expression of the genes, resulting in cancer.

Either a mutation or epigenetic factor may result in cancer by causing a loss or gain of gene function. A loss of gene function may consist of the inactivation of a tumor repressor gene, or inactivation of a gene that programs normal cell death (*apoptosis*), thereby allowing uncontrolled division and growth of a tumor, or inactivation of a gene involved in DNA repair. A gain in gene function may be the increased activation of a gene that normally promotes cell division, resulting in excessive cell division. A **mutagen** is a chemical substance or radiation that causes a gene mutation. An **oncogene** is a mutated gene that causes a cell to become cancerous.

Mutation may alter the cellular environment in ways that facilitate tumor metastases, rather than remaining benign and localized, by affecting the cancer cell's movement or adherence to other cells through

- Increased *angiogenesis* (blood vessel formation) and growth factors, which promote tumor growth into the tumor surroundings
- Altered cell adhesion
- Increased proteolytic enzymes, which increase tumor cell motility and invasiveness
- Avoidance of normal cell death (apoptosis)

Cancer commonly involves multiple gene mutations, which accumulate in life and increase the likelihood of cancer with aging. Some people inherit a tendency toward developing cancer and may develop the cancer earlier than others who do not have the inherited version, since they already have one or more mutations to start. There are more than 50 inherited cancer syndromes (**Figure 6-1**), the most common being breast and lung cancer, followed by colorectal, prostate, skin, and stomach cancer.

Although cancer cells are clones of a predecessor cell, not all the abnormal cells are necessarily the same. This has implications for tumor recurrence, where a treatment optimistically causes a tumor to shrink and regress at first, but not fully. Different cells in the tumor may not be affected by the treatment, or they become resistant to it, and continue to divide.

There are many aspects of cellular interactions apart from gene mutation that have important effects on gene expression and can lead to cancer. These epigenetic influences include:

- *DNA hypomethylation*. Methyl groups that normally attach to DNA molecules generally repress gene transcription. With hypomethylation, the DNA is not repressed and abnormal transcription can occur, e.g. increased cell division. Most tumors show hypomethylation.
- *DNA hypermethylation*. Hypermethylation can contribute to cancer by blocking genes necessary for DNA repair. If DNA cannot be repaired (more than 100 genes are engaged in DNA repair), this increases the possibility of accumulating oncogenic mutations. Hypermethylation (particularly when acting on a promoter region of a gene) can also decrease the transcription of tumor suppressor genes, which control cell growth and division. Hypermethylation can also interfere with the function of genes that control other genes. Environmental factors (such as nutrition, temperature, stress, toxins, drugs, and exercise) can alter DNA methylation.
- *Histones*. DNA wraps around proteins called *histones*, which compress the DNA molecule and assist in determining which genes are turned on or off. The wrapping suppresses the activation of the genes. Some epigenetic factors interfere with histone function and can contribute to carcinogenesis.
- *MicroRNA*. MicroRNAs, by binding to target mRNA molecules, repress translation and act as gene silencers. Overactivity of certain microRNAs may directly reduce the expression of certain DNA repair proteins, thereby allowing for increased mutation and tumor development.

Epigenetic mechanisms of tumorigenesis may be more common than the spontaneous mutation of the gene. They also have greater potential for being reversed.

Common risks for developing cancer include aging, smoking, certain viruses (specifically DNA viruses) and bacteria, exposure to UV light/radiation, toxic chemicals/drugs, hormones, alcohol, decreased exercise, inadequate diet, and obesity. While cancers

FIGURE 6-1. CANCERS WITH A HEREDITARY TENDENCY

- Adrenal gland cancer
- Basal cell nevus syndrome
- Birt-Hogg-Dubé syndrome
- Bloom syndrome
- Bone cancer
- Brain and spinal cord cancers
- Breast cancer
- Brooke-Spiegler syndrome
- Carney-Stratakis syndrome
- Colorectal cancer
- Cowden Syndrome
- Dyskeratosis congenita
- Epidermodysplasia verruciformis
- Epidermolysis bullosa
- Fallopian tube cancer
- Familial adenomatous polyposis
- Familial cylindromatosis
- Fanconi anemia
- Gastric cancer
- Hepatoblastoma
- Hyperparathyroidism, familial
- Kidney cancer
- Li-Fraumeni syndrome
- Medullary thyroid cancer, familial
- Melanoma, hereditary
- Muir-Torre syndrome
- Multiple endocrine neoplasia
- Multiple familial trichoepithelioma
- Neuroblastoma
- Neuroendocrine tumors
- Ocular melanoma and retinoblastoma
- Oculocutaneous albinism
- Oligopolyposis
- Oropharyngeal cancer
- Ovarian cancer
- Pancreatic cancer
- Paraganglioma, hereditary
- Parathyroid gland cancer
- Peutz-Jeghers syndrome
- Pheochromocytoma
- Pituitary gland cancer
- Prostate cancer, hereditary
- Rhabdomyosarcoma
- Rothmund-Thomson syndrome
- Skin cancers
- Small intestine cancer
- Soft tissue sarcoma
- Some types of leukemia and lymphoma
- Stomach cancer
- Testicular cancer
- Thyroid cancer
- Uterine cancer
- Von Hippel-Lindau syndrome
- Werner syndrome
- Xeroderma pigmentosum

can sometimes appear at the site of trauma, there is little evidence that trauma itself causes cancer. Rather, growth factors, cytokines, and angiogenic mechanisms at the site of trauma may provide a more favorable site for metastasizing cancers to settle; or the trauma itself may call attention to a cancer that is already there (e.g. fracture in a bone that is already weakened by a cancer).

Inherited Cancer

Most inherited cancers (**Figure 6-1**) involve defective tumor-suppressor genes. Most such mutations are dominant. When they are recessive, cancer does not develop at first, since this requires the homozygous state. When heterozygous for a recessive oncogene, a cancer does not arise until a second oncogenic mutation occurs later in life in its corresponding normal allele or loss of the function of the normal allele.

Genetic testing is available for some of the hereditary cancers. Some non-hereditary cancers may run in families because of common environmental exposures, e.g. smoking.

A person may inherit a risk for developing multiple tumors. For example, in the *Li-Fraumeni syndrome*, the person inherits a tendency to develop multiple, often rare, tumors, due to a mutation in the p53 gene. Mutations to the BRCA1 (autosomal dominant) and BRCA2 (also autosomal dominant) tumor suppressor genes correspond to a disposition to developing not only breast and ovarian cancers in women, but also are linked to development of pancreatic and prostate cancer, melanoma, and male breast cancer. Both genes are linked to DNA repair. Not everyone gets the disease, just an increased susceptibility. It may require a second mutation later in life to get the disease; hence, the cancer can skip generations, or appear in different organs, depending on which organ gets the second mutation. There are about 2000 different mutations that can occur to BRCA1 and BRCA2 genes. Testing for BRCA1 and BRCA2 may show the person does not have the mutation. However, there is still risk for breast cancer for other reasons common to the general population; or there may be another gene that increases susceptibility; or the test may have missed the BRCA gene.

Mutations to the p53 and Rb1 genes are associated with a variety of diverse cancers. Mutations to the Rb1 tumor suppressor gene is associated not only with retinoblastoma, but sarcomas, melanoma, and brain and nasal cavity cancers. Over 500 kinds of carcinogenic mutations occur in the p53 gene, which is associated with cancers of the lung, skin, head and neck, and esophagus.

Part III. Diagnosis And Treatment of Genetic Disorders

History and Physical Exam

How do you determine if a disease is genetic?
The personal/family history is often telling:

- If the disease is uncommon, is it present in more than one family member?
- Is there a family history of multiple miscarriages or early childhood deaths in more than one family member?
- Is there occurrence of a common disease at an earlier age than expected in more than one family member (e.g. heart disease, cancer)? Of course, if several family members have common back pain, this should not immediately raise concerns that the condition is hereditary, since non-hereditary back pain is so common in the general population.
- Is there a combination of an unusual constellation of symptoms and signs, such as congenital anomalies, mental retardation, and developmental delay?
- Consider relationship to ethnicity, e.g. sickle cell disease in people of African descent; Tay-Sachs disease among Ashkenazi Jews; thalassemia in people of Mediterranean ancestry; hereditary hemochromatosis and cystic fibrosis among people of Northern European descent (**Figure 7-1**). The ethnic-related diseases are generally autosomal recessive.
- Are there outliers such as adoption, extramarital affairs, or consanguinity?

FIGURE 7-1. ETHNIC-RELATED DISEASES

Ethnicity	Disease (Approx. Gene Frequency)
African American	Alpha-thalassemia (~1/30) Beta-thalassemia (~1/75) Cystic fibrosis (~1/65) Sickle cell (~1/14) Spinal muscular atrophy (~1/91)
Asian	Alpha-thalassemia (~1/20) Beta-thalassemia (~1/50) Cystic fibrosis (~1/90) Spinal muscular atrophy (~1/56)
European	Cystic fibrosis (~1/25) Spinal muscular atrophy (~1/47)
French Canadian, Cajun	Tay-Sachs disease (~1/25) Cystic fibrosis (~1/25)
Hispanic	Cystic fibrosis (1/50) Beta-thalassemia (1/40) Spinal muscular atrophy (~1/125)
Jewish Ashkenazic	Canavan disease (~1/49) Cystic fibrosis (~1/28) Familial dysautonomia (~1/40) Gaucher disease (~1/15) Spinal muscular atrophy (~1/46) Tay-Sachs disease (~1/30)
Mediterranean	Alpha-thalassemia (~1/40) Beta-thalassemia (~1/25) Cystic fibrosis (1/29) Sickle cell (1/40)

While a particular genetic disease may have significant genotypic and phenotypic variations in its presentation in the population at large, sometimes the variability in a particular group may be small when the group originated from a small number of people who broke away in the past from the main population and stayed together (termed the **founder effect**).

The physical exam may point to characteristic physical and behavioral findings representative of known inherited diseases. In general, a genetic defect of some kind should be suspected when there is a combination of multiple congenital structural abnormalities (*dysmorphism*) with intellectual disability. About 1-3% of the global population have intellectual disability, about 25% due to a genetic disorder.

Pattern recognition is an important tool in diagnosis. There are extensive databases of genetic diseases in which the practitioner can find diagnostic patterns that fit a patient's history and physical manifestation. For example:

- David Smith's Recognizable Patterns of Human Malformations, Online and Print
- The large OMIM (Online Mendelian Inheritance in Man) database describes conditions due to dominant or recessive gene mutations. There are over 6,000 known inherited diseases. (Medmaster's free downloadable *Atlas of Human Diseases* [Mac, Win] provides rapid access to the OMIM database and information on the Internet in general.)
- In addition to the data and images available on Google and Google Images, YouTube provides many elucidating videos of behaviors and other features associated with a number of genetic disorders.

One of the drawbacks in trying to define a syndrome is that a mutation may cause a single defect that indirectly leads to other problems later in life not directly caused by the mutation itself, such as malnutrition from a deformity that affects eating, or atrophy that results in later life from difficulty walking due to a congenital malformation. These are not strictly part of the syndrome but are secondary effects.

How can you tell if a cancer is hereditary or not? If the child has the tumor but the parent does not, it could be a spontaneous somatic mutation. Or the parent may have the mutation but not express it, especially if it is recessive and both parents have the recessive gene. Or the dominant mutation may have occurred in the gonadal line, in which case the parent wouldn't have it, but the child may, and potentially pass it along to his/her offspring. Suspect a hereditary cancer if:

- An unusual number of people in the family have the same type of cancer (especially if it is an uncommon type).
- The cancer occurs at younger ages than usual (like colon cancer in a 20-year-old), since cancers tend to increase with age with the accumulation of mutations. Be especially suspicious if the same childhood cancer occurs in siblings.
- The person has more than one type of cancer (like a woman with both breast and ovarian cancer). It is also possible, though, that the person had a spontaneous mutation of a gene that is responsible for the development of multiple cancers.
- The cancer occurs in both of a pair of organs (like both eyes, both kidneys, or both breasts). That makes it less likely that one is dealing with an anomalous spontaneous mutation.
- The cancer, while common in the general population, occurs in the sex not usually affected (e.g. breast cancer in a man).

Pedigree Tree

Parents often want to know the likelihood that they, or their children, may inherit a disease found in the family.

The family tree is important to establish that you are really dealing with a hereditary problem and whether it is due to a dominant or recessive gene.

Recall that "Genetic" does not necessarily imply "Hereditary." A mutation in one or more genes can arise after conception, even late in life. For example, only about 5-10% of cancers are hereditary. The rest arise later in life as a mosaic. A **mosaic** is the result of a gene change in a cell after conception. If it occurs right after conception, the defect may affect all cells in the body. If it arises later, it will affect only the part of the body affected by the mutation. The defect could involve a large segment of the body or an area as small as a mole, or in between, e.g. as a tumor in a particular organ. While birthmarks can be inherited, they generally are not. Mosaicism is a reason why "identical" twins, while containing the same DNA, do not appear exactly identical. There may be a mosaic gene change in one of them after conception, or environmental (epigenetic) influences (e.g. diet, temperature, etc.) that affected one and not the other.

Inheritance, as shown in a pedigree chart, may reflect a change that is autosomal dominant, autosomal recessive, X/Y-linked dominant or recessive, or partially dominant or codominant.

- In **codominance** both alleles in a pair of genes are fully expressed and demonstrate traits inherited from both parents; there is a mixture of parental traits. For example, a person heterozygous for blood group alleles A and B expresses both types of antigen. This differs from **partial dominance** (also called **incomplete dominance**), where both traits are blended, rather than separately expressing themselves individually. For example, if a mutation to a gene results in a decrease in the production of a particular protein, the amount of the protein produced from the normal remaining chromosome may suffice to partially drive a needed chemical reaction in the cell, with only a *partial decrease* in function. Or, the decrease in enzyme may be enough to result in a *total loss* in function and be fully dominant.

- There may be *no functional loss at all* with the enzyme decrease if the normal remaining gene alone provides enough enzyme. Cells commonly produce more enzyme than they need and may require only about 5% of it to maintain normality. In that case the condition would be recessive; it would require both genes to be defective to see an adverse phenotypic effect. This is a reason why inborn errors of metabolism are typically recessive conditions. It requires both genes to be abnormal to experience enough decline in enzyme to produce the phenotypic abnormality.
- Depending on additional environmental factors, a person heterozygous for a dysfunctional gene may appear normal, distinctly abnormal, or in-between. A dominant gene may *skip generations*, depending on factors, perhaps environmental, in the skipped generation that modify its expression in the person who carries it. In statistical terms, **penetrance** is the proportion of people in the population carrying the mutation who also express its corresponding phenotypic trait.

Thus, a defective gene that is dominant results in a noticeable abnormality because the defect is sufficient to shut down the normal function, whether or not the gene is heterozygous or homozygous; the remaining normal gene is not enough to sustain normality. A defective gene that is recessive will not result in visible abnormality in the heterozygous state, since the remaining normal gene alone suffices to drive the normal cell function.

A condition that appears recessive (looking at the family tree, neither of the affected patient's parents demonstrates the abnormal phenotype) may actually be dominant if:

- The disease results from a new mutation in the affected individual.
- A parent had a germline mutation (the line of cells that give rise to the gametes), thereby sparing the parent but affecting the offspring.
- Environmental or other epigenetic factors prevent the trait from appearing in the parent and the condition skipped that generation.
- The patient is adopted or the result of an extramarital affair.

The first questions you should ask in interpreting a pedigree:

- Does this look like a dominant or a recessive trait?
- Is it sex-linked or not?
- Determine the probability of the patient's or the patient's offspring acquiring the condition. This calculation may be easy with classic Mendelian (single gene, dominant or recessive) inheritance or may be multifactorial, involving environmental factors or more than one gene, requiring the help of a skilled geneticist to assess the probabilities. Calculation is not always that obvious. For instance, consider **Figure 7-2**. The parents, who are heterozygous for autosomal recessive disease z, which only is expressed in the homozygous state, have a child who does not have the disease. The parents want to know the chances that the child is a carrier. You may think one in two (50%), since two of the four boxes show the heterozygous state. However, it is two in three, since you *know* the child is unaffected by the disease and are now comparing only three boxes, not four. Two out of three of those boxes represent the carrier state.

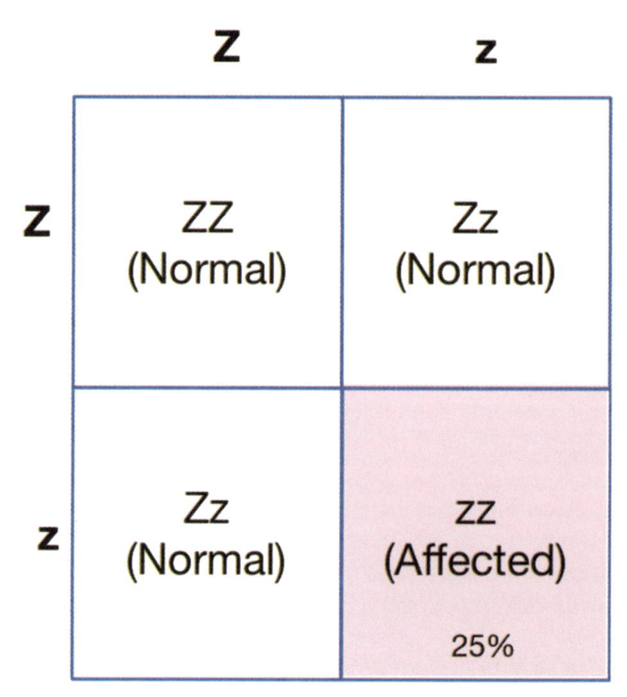

Fig. 7-2.

FIGURE 7-3. SOME HUMAN TERATOGENS

Drugs and Chemicals
- ACE inhibitors
- amphetamines
- alcohol
- aminoglycosides (e.g. gentamicin)
- aminopterin
- androgens
- angiotensin II antagonists
- antithyroid agents
- benzodiazopines
- bromine
- busulfan
- carbamazepine
- chlorobiphenyls
- cocaine
- cortisone
- coumarin
- cyclophosphamide
- danazol
- diethylstilbesterol (DES)
- diphenylhydantion
- etretinate
- excess vitamin A
- heroin
- isotretinoin (Accutane)
- lead
- lithium
- LSD
- marijuana
- mercury
- methimazole
- methotrexate
- penicillamine
- phenobarbitol
- phenytoin
- propylthiouracil
- prostaglandins
- retinoic acid
- streptomycin
- tetracycline
- thalidomide
- toluene
- tobacco
- trimethadione
- valpoic acid
- warfarin

Fever

Ionizing Radiation

Maternal Metabolic Problems
- autoimmune disease
- diabetes
- malnutrition
- phenylketonuria

Microorganisms
- coxsackievirus
- cytomegalovirus
- herpes simplex
- parvovirus
- rubella
- toxoplasma gondii
- treponema pallidum
- Zika virus

Figure 7-2. The chances of having a child that is a carrier is one in two (50%), but the chances that one of the normal children is a carrier is not one in two, but two in three, since you are now comparing three children, not four.

Congenital anomalies can be due to a single gene mutation, a chromosomal structural abnormality, or non-genetic **teratogens** (e.g. infections, chemicals that interfere with fetal development but do not cause DNA mutations and are not inherited), trauma during pregnancy or delivery, or multifactorial causes. Even if a teratogen is discontinued, and there is no gene mutation, the disorder may be too late to correct if the damage has already been done. **Figure 7-3** lists a number of known teratogens. Some of them may be both teratogenic and mutagenic (e.g. radiation, tobacco). Factors that influence their effect are their timing in pregnancy and their duration and dosage. Maximal susceptibility to teratogens occurs between weeks 3 and 8 of pregnancy. Drinking alcohol may cause irreversible damage to the fetus (through *fetal alcohol syndrome*, which causes brain damage and growth problems) at a time when the woman may not even know she is pregnant. Similarly, isotretinoin (retinoic acid, Accutane), a drug commonly used for acne, can cause congenital malformations of the heart, face, thymus, and brain when used in pregnancy, including the first trimester.

A clue to multifactorial inheritance is a clear family tendency, but where the pedigree does not point to a simple Mendelian (one gene) inheritance. Examples of multifactorial diseases include diabetes, asthma, cancer, and mental illness. In Type II diabetes, more than 10 genes are involved in the regulation of glucose, in addition to environmental factors, including diet, obesity, physical activity, and age.

Some genes only make the individual *susceptible* to getting a particular genetic disease, but environmental factors make the difference in the degree to which the disease is manifest or whether it is manifest at all.

In **triallelelic inheritance**, a mutation in three different alleles is required to produce the phenotype. For example, in *Bardet-Biedl syndrome*, a condition marked by obesity, retinitis pigmentosa, and polydactyly (extra fingers), the affected individual must have two defective alleles at one chromosome locus plus an additional defective allele at a different locus.

Some rules on reading family trees (see **Figures 7-4** through **7-11** for examples):

Autosomal dominant (Figure 7-4)

- Multiple generations are affected (a "vertical" pattern).
- Heterozygotes show the disease.
- With a relatively rare disease, the patient has only one affected parent.
- Males and females are equally likely to inherit the disease.
- Only one affected parent is necessary to pass the disease.
- The offspring of an affected person has a 50% chance of acquiring the disease.
- Malformations are often autosomal dominant, unlike errors of inborn metabolism, which typically are autosomal recessive.

Autosomal recessive (Figure 7-5)

- The parents of the patient do not have the condition, but there may be many siblings affected in a given generation (a "horizontal" pattern), with none in subsequent generations, since the condition is recessive. Each parent is heterozygous for the abnormal recessive gene.
- The siblings of an affected child have a 25% chance of acquiring the disease, a 50% chance of being a carrier, and a 25% chance of not carrying the abnormal gene at all.
- Males and females are equally likely to inherit the disease.
- More than one sibling may be affected, suggesting that the disease was not due to a spontaneous mutation in the patient.
- When the disease is rare, the chances are increased that the parents may be close relatives.
- When two people with the disease mate, 100% of the offspring will have the disease, since both parents are homozygous recessive for the gene.
- When a person with the disease mates with someone who is normal and is not a carrier, none the offspring will have the disease, but all will be carriers.

X-linked dominant (Figure 7-6)

- The appearance of the pedigree resembles that of an autosomal dominant disorder, except that all daughters, but no sons, of an affected *father* will have the disorder. That is because all sons of the affected father inherit only his Y chromosome. His daughters will definitely inherit his defective X chromosome and manifest the disease. If it is the *mother* who has the disease, you would on average expect 50% of the daughters and sons to have it, and none of the sons or daughters will be carriers if they don't have the disease.
- The condition is often less pronounced in females, because they have a normal second X chromosome to help balance the defective one.

X-linked recessive (Figure 7-7)

- No daughters of an affected (homozygous recessive) mother will acquire the disease, since they only inherit one X chromosome from their mother. Those daughters, though, will all be carriers.
- All sons of an affected homozygous mother will acquire the disease. If the mother is heterozygous for the condition, 50% of daughters will be carriers and 50% of sons will have the condition.
- All the daughters of an affected father will be carriers, but none of the father's sons will be carriers or have the disease, since sons only inherit their father's Y chromosome.
- Commonly, neither parent of an affected daughter will manifest the disease, since the child inherited one defective recessive gene from each parent, each of whom is heterozygous and does not manifest the disease. The mother of an affected male child would be the parent who passed down the recessive gene, since the father only passes down the Y chromosome.

Inborn errors of metabolism (IEM) tend to be autosomal recessive mutations. One reason may be that one normal gene alone suffices to supply enough normal enzyme to regulate the metabolic reaction and prevent the disease. You need the recessive state to reduce the supply of effective enzyme enough to cause a clinical problem. Although rare, there are some 500 different diseases of IEM.

Inherited immunodeficiency diseases also tend to be recessive. Many of the mutations are on the X chromosome, thereby calling the relatively rare diseases to greater attention, since they become manifest in males with only one defective gene.

Y-linked (Figure 7-8)

- All sons of an affected father will have the condition, since the Y chromosome is handed down from male to male. Mutations may result in abnormalities of sexual development.

- No female can acquire the condition, since it is Y-linked.
- In **pseudoautosomal inheritance**, the gene of interest is paired and lies on both the Y and X chromosome, in the small homologous chromosomal region of X and Y where there are corresponding alleles. The inheritance pattern then behaves like an autosomal one. Examples of pseudoautosomal inheritance are *Y-linked Hodgkin disease, Langer mesomelic dysplasia*, and *dyschondrosteosis*.

Mitochondrial (Figure 7-9)

- When mitochondrial dysfunction is the result of a nuclear DNA mutation that affects the mitochondria, the inheritance can be typically Mendelian autosomal dominant, autosomal recessive, or X-linked. If the DNA mutation is in the mitochondria, there can be no dominance or recessiveness, since mitochondrial DNA is not paired. If originating in the mitochondria:

  a. Males and females will inherit the disease, but only from the mother, since sperm do not contribute mitochondria.

  b. Females with the disease will pass on the mutation, but males will not, since sperm cells do not pass on mitochondria.

The degree to which a mitochondrial disease is expressed may vary considerably since there are about 1000-2000 mitochondria in a cell, not all of which may be affected, and many cells may not be affected. The proportion of mutant mitochondrial DNA will determine the degree of expression and penetrance. Mitochondrial diseases often affect muscle, brain, and nerve cells more severely than other body areas because these cells require more energy, and mitochondria are a key source of energy in the form of adenosine triphosphate (ATP).

Figure 7-4. Example of autosomal dominant inheritance. Vertical pattern of involvement. (Circle = female; square = male; filled-in icon = affected by disease)

Figure 7-5. Example of autosomal recessive inheritance. Horizontal pattern of involvement.

Figure 7-6. Example of X-linked dominant inheritance. Male cannot pass the disease to sons.

Figure 7-7. Example of X-linked recessive inheritance. Transmitted only to sons of an affected mother.

Figure 7-8. Example of Y-linked inheritance. Affects only males.

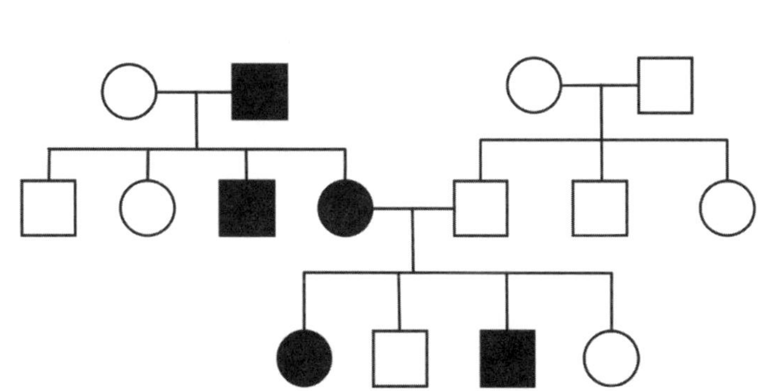

Fig. 7-4.

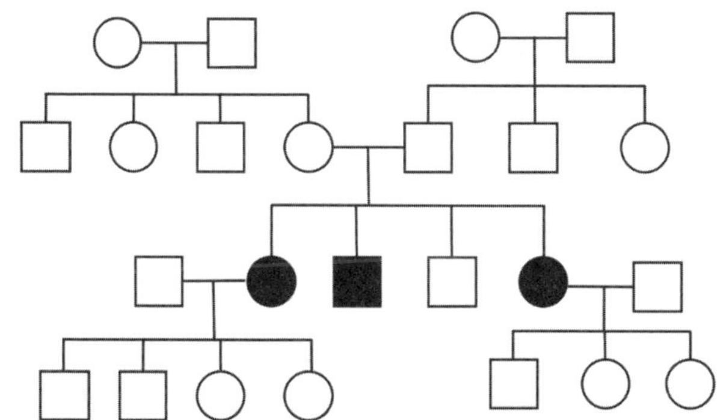

Fig. 7-5.

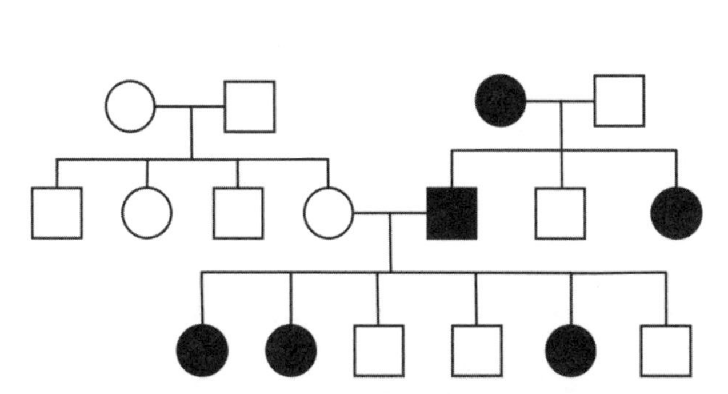

Fig. 7-6.

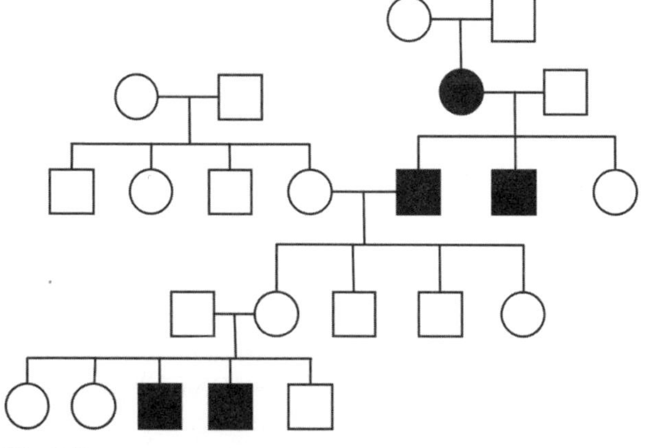

Fig. 7-7.

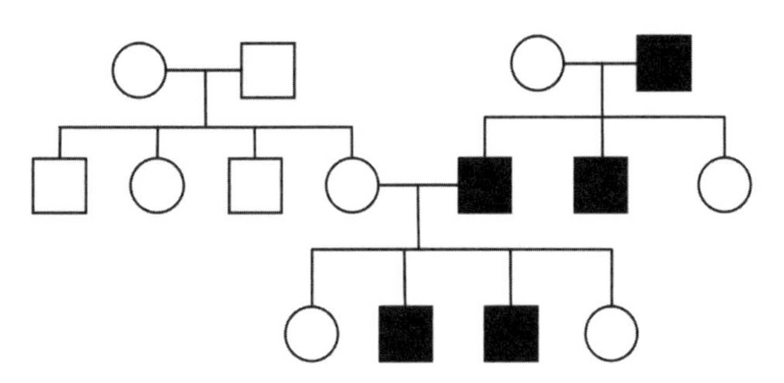
Fig. 7-8.

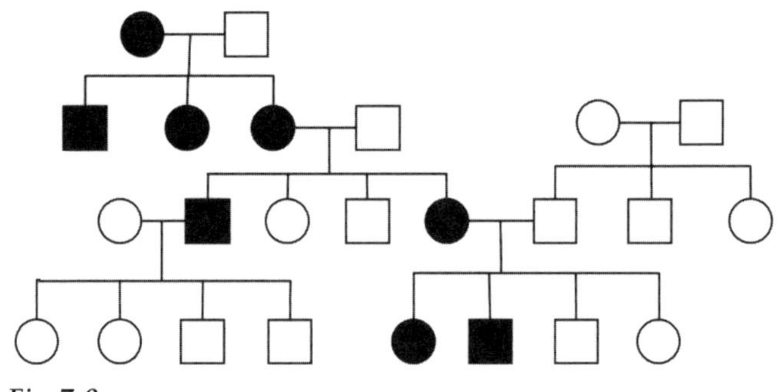
Fig. 7-9.

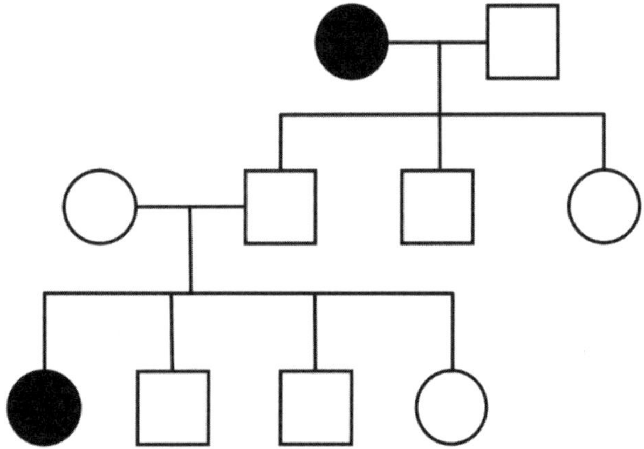
Fig. 7-10.

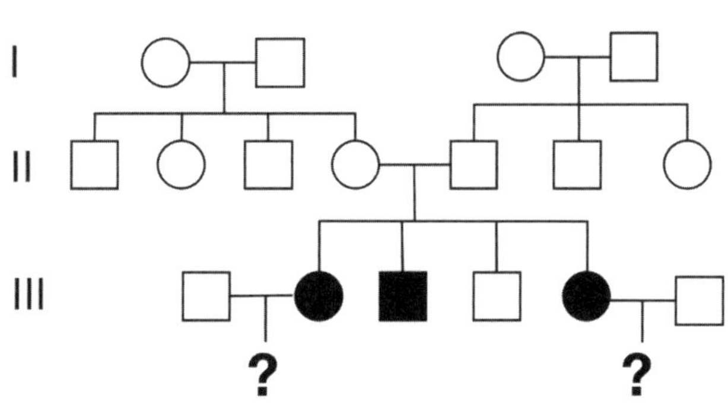

Fig. 7-11.

Figure 7-9. Example of mitochondrial-linked inheritance. Only the mother can pass on the disease.

Figure 7-10. Example of skipped generation for a dominant gene. Skipping may occur if environmental or additional genetic factors prevent the trait from appearing in the parent's generation, and the trait then skips to the next generation.

Figure 7-11. Recessive or dominant? This tree resembles the tree in **Figure 7-5** for autosomal recessive except for the missing last generation line. In this case, each of the affected potential parents in generation III wants to know if there is a risk that their future children may express this rare disease. If the disease is recessive, there is little risk, since it is a rare recessive gene that will not likely be expressed in generation IV. However, one of the parents in generation II could be dominant, but with the defective gene arising only in that parent's germline (sperm or ovum), and not in the rest of the parent's cells. Thus, the parent in generation II would not be expected to exhibit the trait. However, the offspring of generation III would have a 50% chance of inheriting and exhibiting the trait, since one parent in generation III is dominant for the disease. The accuracy of a pedigree diagnosis often rests on having a sufficient number of people in the tree for whom medical information is known.

- A significant congenital anomaly occurs in about 3-4% of living births.
- The prevalence of a defective gene in the population depends not only on the frequency

FIGURE 7-12. EXAMPLES OF HETEROZYGOUS ADVANTAGE	
Disease	**Protects Against**
Sickle cell/alpha- and beta-thalassemia	Malaria
Cystic fibrosis	Cholera, typhoid, Tb
Triosephosphate isomerase deficiency	Damage from harmful free radicals to proteins, lipids, and DNA
Smith-Lemli-Opitz syndrome	Rickets through increased vitamin D synthesis

of the mutation, but in part whether it is dominant or recessive. One would expect life-threatening dominant mutations to die out in the population faster than recessive mutations, since an abnormal recessive gene may persist in the heterozygous state through many generations, without being noticed. However, this rule is less true for dominant mutations whose phenotype is not life-threatening until later in life beyond the reproductive age (e.g. Huntington disease). The gene may then be passed during the person's reproductive life and seen even more frequently than a recessive trait, which requires the homozygous state to be expressed.

- Recessive genes appear more often than expected in the homozygous state due to a close bloodline relationship.
- Also, some genes may hang around longer because of **heterozygous advantage**. I.e., despite the harmful effect of the gene, it may have a beneficial effect. For instance, the recessive gene for sickle cell anemia helps to protect against malaria; the recessive gene for cystic fibrosis helps protect against cholera by decreasing the amount of dehydration (**Figure 7-12**).

Mutations per se are not necessarily bad. Favorable mutations are needed for evolution to progress.

8

Laboratory

Prenatal and newborn screening is standard in trying to detect in advance a number of genetic diseases that can be managed with early detection. Carrier testing helps in establishing the tendency to inherit a particular genetic disease and is particularly useful in family planning.

Should You Do Genetic Testing?

Some kinds of genetic testing may be very beneficial in planning treatment. Other tests are merely informative, when there is no known treatment, but can help the family and patient psychologically just to know, as well as to engage in family planning. Some patients and families may desire genetic diagnostic information while others do not. In some cases it may even be psychologically beneficial not to know about something for which there is no known cure, especially if it is a condition not likely to appear until later in life or is not likely to be inherited. In addition, often you do not even need to know the underlying area of fault in the DNA sequence to treat the symptoms and signs, so it may not be necessary to go through the testing.

Counsel before initiating genetic testing, and explore the family's concern about their or their children's acquiring the disease. The patient should understand the pros and cons of the diagnostic test and participate in the decision-making of whether or not to undergo testing in the first place, considering the cost, practical considerations in doing the test, and effectiveness of treatment. As a general rule, don't do a diagnostic test unless the results may prove useful (e.g. lead to a change in lifestyle, help family planning, provide evidence that the disease is starting to appear so that treatment measures can be instituted early, point to a specific drug or other treatment modality, or provide peace of mind in knowing). The benefits should outweigh the physical and psychological harm to the person being screened.

The results should also be kept confidential. It can be significantly disturbing for a patient to fear that the test results may be held against him/her in trying to obtain medical insurance or employment, or become known in his/her social network.

Indications for Referral and Cytogenetic Testing

About 3-4% of newborns have some kind of birth defect. In general, consider referral if you suspect that the patient has a genetic disorder or is at risk for one. The specialist can recommend the appropriate tests in view of the family history and physical findings, indicate prognosis, suggest treatment, and refer to other specialists (e.g. cardiologist, surgeon, ophthalmologist, social worker, nutritionist, physical therapist), depending on the problem.

What are the indications for cytogenetic testing?

- The family history suggests a genetic disease (e.g. other members of the family are affected similarly with an unusual condition, or, when the condition is more common, have a history of early deaths, or have a known chromosomal abnormality).

- If a couple, e.g. for premarital information, requests testing and is in an ethnic group where certain conditions are common, especially if the couple are closely related (**Figure 7-1**)
- Advanced maternal age (> 35 years)
- The child has a worrisome developmental delay, congenital defect, intellectual impairment, neurological problems (e.g. seizures, hypotonia, ataxia), or striking behavioral problems (e.g. autism, psychiatric disorder).
- Prenatal or newborn screening suggests an abnormality suggestive of a genetic origin.
- A woman who has multiple miscarriages, infertility, or early infant death
- Ultrasound of the fetus suggests a significant abnormality.
- Amenorrhea or premature menopause, hypogonadism, or infertility
- Certain cancers (e.g. leukemia and other tumors) where the particular genomic abnormality may guide therapy

Cytogenetic Screening Tests

Genetic screening tests can be done on a blood sample, hair, skin, amniotic fluid, or other tissue. Newborn screening is done by a prick of the baby's heel. Testing is not confined to looking at DNA or RNA; it also examines proteins and other metabolites that are abnormal in certain genetic diseases and may provide the basis for the diagnosis and treatment.

- **Diagnostic testing** tries to determine the cause of the condition. The test will depend on the particular genetic condition that you are concerned about.
- **Carrier testing** tries to determine whether a person with a normal phenotype may be a carrier of the disease. In general, carrier testing on an individual needs to be done only once in their lifetime, but newer screening panels may be recommended with advances in screening for other related mutations. Ideally, carrier testing should be done before pregnancy and part of marital or premarital planning. If the person being tested is found positive for the condition, the partner, and significant relatives as well, should be given the option of undergoing testing, too. If both partners carry the defective gene, offer genetic counseling. Of course, both individuals have the choice of whether or not to do the screening, and the information should be kept confidential. The family history and ethnic background are important determinants of which tests to perform. Genetic carrier tests commonly recommended are those for
    - cystic fibrosis
    - sickle cell disease
    - the thalassemias
    - fragile X syndrome
    - spinal muscular atrophy

Screening for other conditions may be recommended, depending on family history, if other specific diseases are known or suspected. The ethnic group of the individual should be considered in planning what to screen for, since certain diseases are significantly more common in certain ethnic groups (**Figure 7-1**). For instance, in the Ashkenazic community (people of Eastern and Central European Jewish descent), screening is particularly relevant for Tay-Sachs disease, familial dysautonomia, cystic fibrosis, and Canavan disease in view of their high carrier frequency. Some geneticists recommend a broader screening in this group for Bloom syndrome, Gaucher disease, Niemann-Pick disease, Usher syndrome, familial hyperinsulinism, von Gierke disease, Joubert syndrome, maple syrup urine disease, and mucolipidosis IV.

- **Prenatal testing**, apart from testing the pregnant mother's general health, aims to discover in advance whether the fetus may have the disease. In addition to ultrasound, diagnosis can be done by chorionic villus biopsy or amniocentesis, both of which carry a small risk (<1%) of miscarriage. Alternatively, you can examine fetal DNA or cells in the maternal blood, which often is possible in pregnancy. Commonly, what prompts the testing is an abnormality found on imaging tests of the fetus, such as ultrasound or MRI. Ultrasound, for instance, may suggest the appearance of Down syndrome, which should be considered together with a maternal blood test to see if there are certain proteins in the mother's blood that correlate with the condition. Prenatal screening may be suggested when the mother is more than 35 yrs old; one parent is known to carry a chromosomal rearrangement; both parents may carry a potentially harmful recessive gene, for which 25% of newborns will be affected (e.g. Tay-Sachs, cystic fibrosis); the mother had a previous pregnancy with a child with Down syndrome; or the mother carries a harmful X-linked gene, for which 50% of the male children will be affected (e.g. muscular dystrophy).
- **Preimplantation testing** can be done on an early embryo prior to implantation to determine if the embryo is carrying a defective gene. At the earliest stages of development, a single cell can be removed for analysis without harming embryonic development. Preimplantation testing can test for more than 100 genetic diseases.

- **Predictive testing** (**Presymptomatic testing**) is done to determine whether an asymptomatic person carries a particular mutation, and the risk that they will develop the disease, particularly when there is already a family member with the disease. For instance, someone with a BRCA1 gene mutation will have a 65% risk of eventually developing breast cancer.
- **Neonatal screening tests** (**Figure 8-1**) are for early disease detection in the newborn, when initiating treatment is most effective. Although many of the conditions are rare, they are included because of the cost-effectiveness of the testing and usefulness of early intervention. Treatments include nutritional intervention to avoid ingesting harmful substances (e.g. phenylalanine in phenylketonuria), provide missing substances (e.g. hormone replacement), surgery (e.g. cochlear implant for deafness), and careful attendance to environmental exposures (e.g. sickle cell trait). The recommended screening profile may vary by state.
- **Infertility Workup**. An infertility workup helps determine the cause of infertility and whether the problem originates with the woman or man.

1. *The woman*: Is she ovulating? This is when the ovary is releasing an egg, usually about day 14 of the menstrual cycle. Are there symptoms of ovulation such as abdominal cramps, change in body temperature, change in vaginal discharge, the level of luteinizing hormone, which can be determined in an ovulation predictor kit? If menstrual cycles are irregular, further information can be obtained from Follicle Stimulating Hormone (FSH), estrogen, progesterone, and luteinizing hormone (LH) blood levels, and imaging tests to examine whether ovarian follicles, uterus and Fallopian tubes are normal. AMH (anti-Mullerian hormone) level can help assess the woman's egg supply. A hysterosalpingogram is an x-ray of the uterus and Fallopian tubes to assess their normality. Is there evidence of a pelvic inflammatory disease (e.g. gonorrhea, chlamydia)?
2. *The man*: A semen analysis that assesses sperm count and motility helps determine whether the problem lies with the male. Potential problem areas to investigate include urinary or prostatic infection, undescended testes, varicocele, retrograde ejaculation, low serum testosterone, FSH, LH or prolactin hormone levels, anti-sperm antibodies, Y chromosome or other chromosome mutation (as in cystic fibrosis, where there may be absence of the vas deferens).

- **Karyotyping**. Chromosome analysis is used in newborns with congenital malformations or older patients with mental dysfunction and other suspected genetic abnormalities. Chromosomal analysis is also used as part of the workup for multiple miscarriages and infertility. While karyotyping may pick up major visible changes in the chromosomes, specialized tests are more helpful in picking up smaller changes in the gene sequence. Gene sequencing searches for the specific gene mutations in the genome. With the advancing field of whole genome sequencing, a number of older tests are becoming less used. Here, we mention the currently used PCR, FISH, and Next Generation Sequencing.

Polymerase Chain Reaction (PCR)

The **Polymerase Chain Reaction (PCR)** is a way of reproducing many identical amounts of a small sample of DNA (sometimes RNA) in quantities large enough to test accurately by comparing the sample with a known source, thereby enabling the laboratory to identify the sample. The comparison is often done by placing the sample and known source together in an electrophoretic gel to confirm that the sample and source migrate the same way in the gel and have the same nucleotide sequences. PCR is used in forensic investigations to compare the DNA found on a victim with that of the suspect or a database of potential suspects. Clinically, it is also used to assess genetic paternity, detect pathogenic organisms, and identify specific genetic mutations.

The variable number of **tandem repeats (VNTRs, minisatellites)** are useful in genetic testing even while they may not have a known function. VNTRs can be picked up on gel electrophoresis, where they have different migration rates, and can be used in DNA fingerprinting for forensic analysis and paternity testing.

Fluorescence In Situ Hybridization (FISH)

FISH (Fluorescence In Situ Hybridization) is a way to visualize and map specific genes or parts of genes. A fluorescent probe, of which there are many kinds, containing known short nucleotide sequences of DNA (or RNA), is applied to a preparation of the patient's chromosomes. The probe is designed to target and adhere to its specific complementary DNA sequence in the patient's chromosomes. This enables the visualization of whether there is a deletion or duplication of genes, a translocation or abnormal number of chromosomes, chromosomal rearrangements, or significant mutations that do not allow the probe to adhere. A number of fluorescent probes of different colors can be applied at

FIGURE 8-1. RECOMMENDED NEWBORN SCREENING (AMERICAN COLLEGE OF MEDICAL GENETICS)

Disease/Frequency	Treatment
Argininosuccinic aciduria (< 1 in 100,000)	High calorie, protein-restrictive diet; arginine supplementation; possible dialysis
3-Methylcrotonyl-CoA carboxylase deficiency (> 1 in 75,000)	Low leucine diet, as body cannot break down the amino acid, leucine
Beta-ketothiolase deficiency (< 1 in 100,000)	Intravenous fluids, glucose, electrolytes, and bicarbonate
Biotinidase deficiency (> 1 in 75,000)	Biotin
Carnitine uptake defect (< 1 in 100,000)	L-carnitine supplements
Citrullinemia (< 1 in 100,000)	Removal of ammonia from body, dialysis, low-protein diet; liver transplantation in some cases
Classical galactosemia (> 1 in 50,000)	Low galactose diet
Congenital adrenal hyperplasia (> 1 in 25,000)	Steroids
Congenital deafness (> 1 in 5,000)	Cochlear implants
Congenital hypothyroidism (> 1 in 5,000)	Thyroid hormone replacement
Critical congenital heart disease	Medication, surgery, heart transplant depending on severity
Cystic fibrosis (> 1 in 5,000)	Antibiotics, mucus-thinning medication, airway clearance therapy
Glutaric acidemia Type I (> 1 in 75,000)	Low lysine diet, carnitine supplementation
Hb S/beta-thalassemia (> 1 in 50,000)	Blood transfusions; supportive, including drinking plenty of water, avoiding climate extremes; hydroxyurea
Homocystinuria (< 1 in 100,000)	Vitamin B6, low protein/methionine diet, betaine, folate/ cobalamin supplementation
Hydroxymethylglutaryl lyase deficiency (< 1 in 100,000)	L-carnitine dietary supplementation
Isovaleric acidemia (< 1 in 100,000)	Dietary protein restriction, especially leucine; glycine, L-carnitine; avoid alcohol; treat hyperammonemia with sodium phenylacetate and sodium benzoate
Long-chain hydroxyacyl-CoA dehydrogenase deficiency (> 1 in 75,000)	Low-fat diet; medium-chain triglyceride oil
Maple syrup urine disease (< 1 in 100,000)	Dietary restriction of branched-chain amino acids
Medium-chain acyl-CoA dehydrogenase deficiency (> 1 in 25,000)	Maintain blood sugar levels with complex carbohydrates and frequent feeding; avoid prolonged fasting
Methylmalonic aciduria, cblA and cblB forms (< 1 in 100,000)	Low protein diet, L-carnitine, antibiotics, ammonia detoxification, sometimes liver and kidney transplantation
Methylmalonyl-CoA mutase deficiency (> 1 in 75,000)	Low protein diet, sometimes liver and kidney transplant
Multiple-CoA carboxylase deficiency (< 1 in 100,000)	Oral biotin
Phenylketonuria (> 1 in 25,000)	Limit dietary phenylalanine
Propionic acidemia (> 1 in 75,000)	Protein restriction; carnitine; sometimes liver transplant; physical/ occupational therapy
Sickle cell anemia (> 1 in 5,000); (among African-Americans 1 in 400; 1 in 12 African-Americans in the US have sickle cell trait)	Blood transfusion; avoid pain episodes (drink lots of water, avoid temperature extremes, avoid high altitudes and low oxygen); hydroxyurea; stem cell transplant
Trifunctional protein deficiency (< 1 in 100,000)	Low long-chain fatty acid/high carbohydrate diet; L-carnitine
Tyrosinemia I (< 1 in 100,000)	Dietary restriction of tyrosine and phenylalanine; nitisinone (inhibitor of 4-hydroxyphenylpyruvate dioxygenase)
Very-long-chain acyl-CoA dehydrogenase deficiency (> 1 in 75,000)	Low-fat/high-carbohydrate diet, frequent feeding

once, thereby allowing the analysis of the status of a number of different chromosome sites at the same time. One advantage of FISH over microscopic karyotyping is that it can be done on nondividing cells, in addition to exploring smaller parts of the chromosome.

Next Generation Sequencing

Next Generation Sequencing involves the actual sequencing of either the entire genome, which contains some 3 billion base pairs, or just the entire set of all the exons (the protein-coding segments of DNA), which are only about 1-2% of the entire genome, but where most of the genetic diseases occur. With modern advances, the entire genome can now be sequenced rapidly at reasonable cost. Even a single cell can be gene-sequenced.

Problems with Next Generation Sequencing:

- While 99.9% of human DNA is the same, no two normal people are completely alike genetically (even identical twins have been found to have differences in gene copy numbers). Variations in some nucleotide base pairs amongst some 3 billion base pairs are common, normal, and nothing to worry about. A single pathological mutation may be hidden and difficult to find, and an unusual difference in a gene sequence may be of no consequence. Genome sequencing is more valuable when there is a known pathological nucleotide sequence being sought, especially one that appears in a number of affected family members.
- The environment may play a significant role in altering the phenotype without altering the DNA sequence (epigenetic influences). There then will be no abnormality in the sequencing.
- Sometimes a number of different mutations, involving different genes, can result in the same disease, making it harder to determine the cause of the disease.

Despite these difficulties, Next Generation Sequencing will continue to increase in value as a screening and diagnostic test. With everyone genotyped, including their nuclear DNA and mitochondrial DNA, nucleotide sequences that were originally thought to be pathological may be found to be normal variants; and pathological nucleotide sequences will be easier to identify. This will lend greater meaning to the lab report, when it comes back as either:

a. Normal sequence
b. A variant that is known to be pathogenic, benign or of unknown significance.

9

Treatment of Genetic Disorders

Once a genetic disease is diagnosed, what can you do about it?

Genetic diseases are often treatable. The treatment of genetic disorders may include dietary and other environmental changes, surgery (e.g. to correct a congenital malformation, or bone marrow transplantation to correct an anemia), enzyme replacement, detoxification, drugs and radiation therapy for cancer, gene therapy (which interacts directly with the genes), physical/occupational therapy, and counseling. While cofactors, such as metallic ions and vitamins, are not proteins, they bind to the enzymes and affect their function, and can be useful supplements to enhance the action of enzymes.

- If there is an environmental (epigenetic) cause to genetic malfunction, you may be able to alter the environment to reverse or prevent the adverse influence. Changing an **epigenome** (the sum total of chemical changes to DNA and histones that do not change the DNA nucleotide sequence) would be easier than changing a genome (the DNA nucleotide sequence itself). Even if a condition is not epigenetic in origin, improving the environment can often be helpful (e.g. in sickle cell trait, avoiding low pressure in an airplane, dehydration, and intense physical activity in sickle cell trait).
- Inborn errors of metabolism, which generally are caused by hereditary enzyme defects, often have treatments. Some may include removing an accumulating toxic factor through diet or medication, or supplying a missing and needed metabolic product. For instance, *phenylketonuria*, a hereditary defect in the enzyme that changes phenylalanine to tyrosine, results in the accumulation of toxic amounts of phenylalanine. Treatment includes reduction of dietary phenylalanine. In *familial hypercholesterolemia*, a chromosomal defect results in the inability to remove low-density lipoprotein cholesterol. Treatment includes dietary changes and lipid-lowering agents.
- In some cases the enzyme defect may result in the decreased production of a needed product that has to be replaced. For instance, *21-hydroxylase deficiency*, the most common hereditary enzyme defect in steroid biosynthesis, results in decreased synthesis of corticosterone, with adrenal hyperplasia. Treatment includes administering corticosteroids.
- While most hereditary enzyme defects, such as phenyketonuria and 21-hydroxlate deficiency, are *loss-of-function mutations*, where the needed enzyme is absent or ineffective, in some cases the problem is a *gain-of-function mutation*, where a new, mutated enzyme has a different or enhanced function. For instance, in *phosphoribosylpyrophosphate (PRPP) superactivity*, there is a mutant hyperactive PRPP synthetase enzyme, which leads to uric acid overproduction and accumulation, and gout. It is treated with uric acid lowering agents. In *hereditary hyperthyroidism*, there is an abnormality in the

hormone for the Thyroid Stimulating Hormone (TSH) receptor on the thyroid cell, making the thyroid cell more active in producing thyroid hormone. The condition can be treated with anti-thyroid medication or surgery.

- In other cases (e.g. *Gaucher disease*, *Fabry disease*, and *Pompe disease*), it is possible to replace the missing enzyme.
- Family counseling is an important aspect of therapy. Knowing the genetic cause of a condition and its hereditary pattern can provide psychological assurance and advice for family planning, as well as suggest the best treatment plan.
- Diseases that are worsened by environmental factors (e.g. alcohol and other substance abuse, smoking, ultraviolet radiation, obesity, herbicides, and pesticides) can be approached clinically by trying to adjust the environmental influences.
- Epigenetic treatments that target DNA methylation/demethylation and histone acetylation/deacetylation hold promise for treating genetic diseases. When epigenetic factors cause cancer, there may be a greater chance of reversing the condition than if the cancer is due to mutation. Inhibiting methylation with methyltransferase inhibitors, for instance, can restore the expression of tumor repressor genes silenced by methylation.
  a. *Histone deacetylase (HDAC) inhibitors*, by inhibiting histone deacetylases, cause the hyperacetylation of histones, and allow DNA to wrap less tightly around the histones, thereby increasing gene expression and apoptosis (normal cell death). They have been used in the treatment of various cancers, to promote cell cycle arrest and apoptosis and inhibit the proliferation of tumor cells and new blood vessels.
  b. *Histone acetyltransferase (HAT) inhibitors*, by inhibiting histone acetylase, cause the deacetylation of histones and can decrease gene expression. They are undergoing trials in certain cancers, Alzheimer disease, diabetes, and hyperlipidemia.
  c. *Histone methyltransferase (HMT) inhibitors* inhibit histone methylation and gene overexpression, as may occur in certain cancers, where there is overexpression of cell division.
  d. *DNA methyltransferase (DNMT) inhibitors* cause demethylation of genes, thereby increasing their expression. They can be used, for instance, to increase the expression of gamma-globulin genes by hypomethylating them, and increase the production of fetal hemoglobin in patients with sickle cell disease.
  e. *HDM-2 inhibitors* block the harmful binding of HDM2 protein to the tumor suppressor protein p53, thereby helping control the growth and spread of cancer.

One of the cautions that must be considered in epigenetic drug treatment is its specificity. You would like to have a treatment that is specific for the disease, such as vitamin C for scurvy, rather than generalized, where there may be unprecedented side effects. Scurvy results from severe deficiency of vitamin C in the diet. This leads to an epigenetic hypermethylation of DNA, which inhibits the production of collagen in the skin and blood vessels, leading, among other things, to gum disease, bleeding, and poor wound healing. The condition can be treated by administering vitamin C, which in effect is an epigenetic treatment.

- Gene therapy, directly changing the genes themselves, is a rapidly growing research field (see CRISPR below).

CRISPR

CRISPR (**C**lustered **R**egularly-**I**nterspaced **S**hort **P**alindromic **R**epeats) gene alteration therapy holds great potential for curing many human genetic diseases. It also has other diverse uses, such as enhancing food crop nutritional value and resistance to drought and plant diseases. It can also be used to make animals more vigorous, e.g. in developing more bulky muscles in cattle, or making their organs more human-like in composition and function, so that they can be used more effectively for organ transplantation; or enabling animals to produce human proteins for treatment; or rendering mosquitoes incapable of carrying malaria; and engineering microbial genomes for enhanced fuel production (**Figures 9-1** and **9-2**).

Figure 9-1. CRISPR. A segment of RNA (guide RNA) that identifies and targets the area on the gene of interest is attached to a Cas9 enzyme, which acts as a gene cutter, slicing out the defective gene segment ("cut and remove") and thereby inactivating the harmful gene. Alternatively, the gene can be converted to a normal gene by inserting a normal gene segment that replaces the defective segment ("cut and replace")(see also **Figure 9-2**).

Figure 9-2. CRISPR/Cas9 and its potential uses. PAM = **P**rotospacer **A**djacent **M**otif, a 2-6 base-pair sequence needed for Cas9 to bind and cut the target DNA.

The current ease and relatively low expense of decoding a human genome have greatly increased the therapy potential of CRISPR/Cas9. It is now possible to map exactly where defective genes lie on

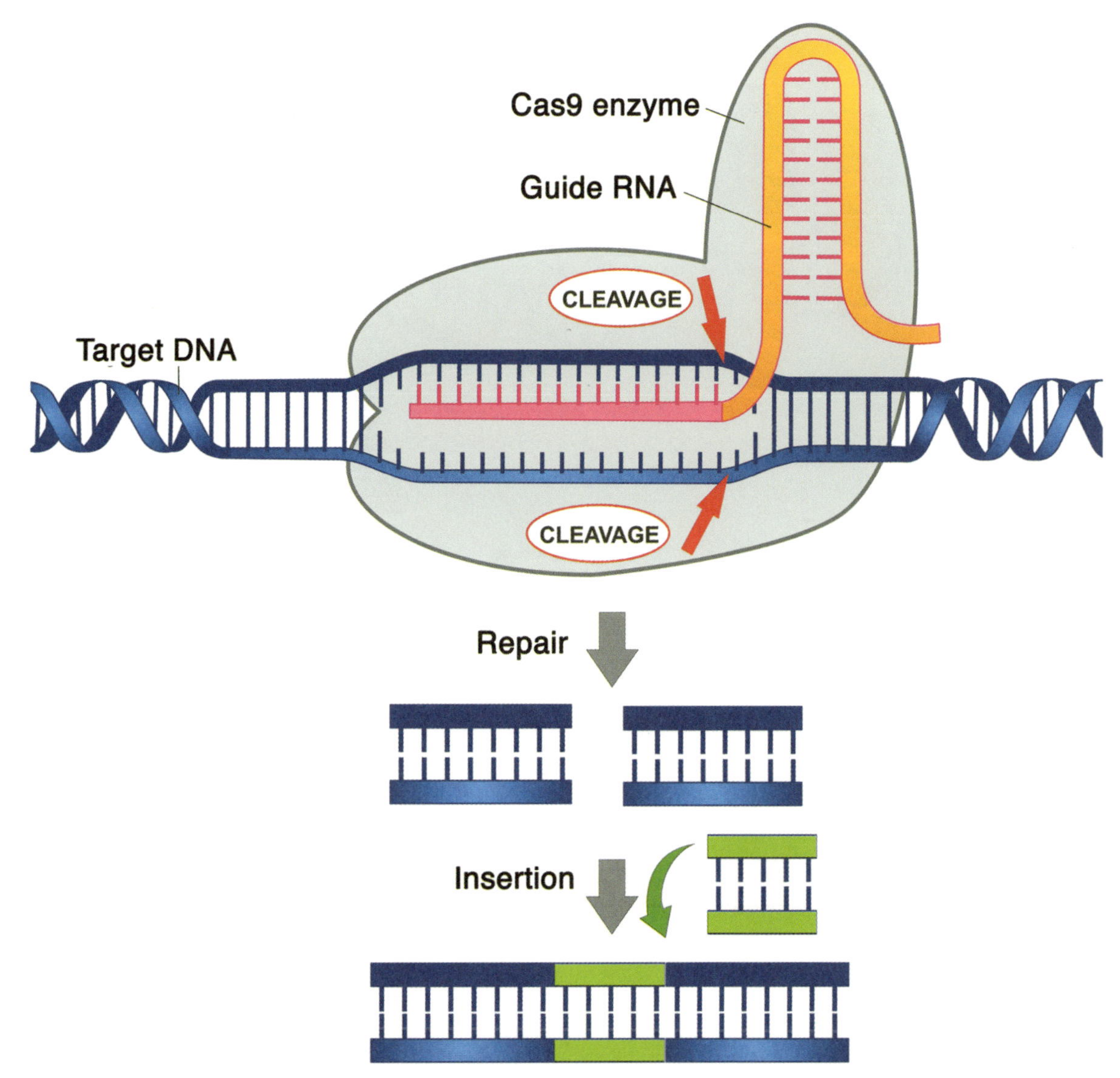

Fig. 9-1.

the chromosome and use CRISPR to directly treat defective genes with the following advantages:

1. Where the defect is a single nucleotide mutation, CRISPR can remove the mutated nucleotide or replace it (or a length of the mutated gene) with a normal sequence. **Figures 9-1** and **9-2** show the general way this is done. When a gene mutation produces an abnormal harmful protein, using CRISPR to simply disrupt the gene with a small insertion or deletion can be enough to disable the gene (called a **knockout**), which will help correct the deficit.

When a disease is due to a defective dominant gene in the heterozygous state, CRISPR can be used to disable the dominant gene, in which case the remaining normal recessive gene may be sufficient to reverse the disease. In the case of a recessive homozygous disorder, CRISPR can be used to disrupt the defective gene and substitute a normal gene segment.

2. Genetic changes produced by CRISPR are lifelong, rather than the metabolic changes produced by drugs that have to be continually administered.
3. Sometimes cancer or other diseases may result from the combined influence of a number of genes. Knocking out one or more of those genes may interfere with their composite action, which has the potential of successfully treating certain cancers.

Presently, though, CRISPR therapy has a number of challenges:

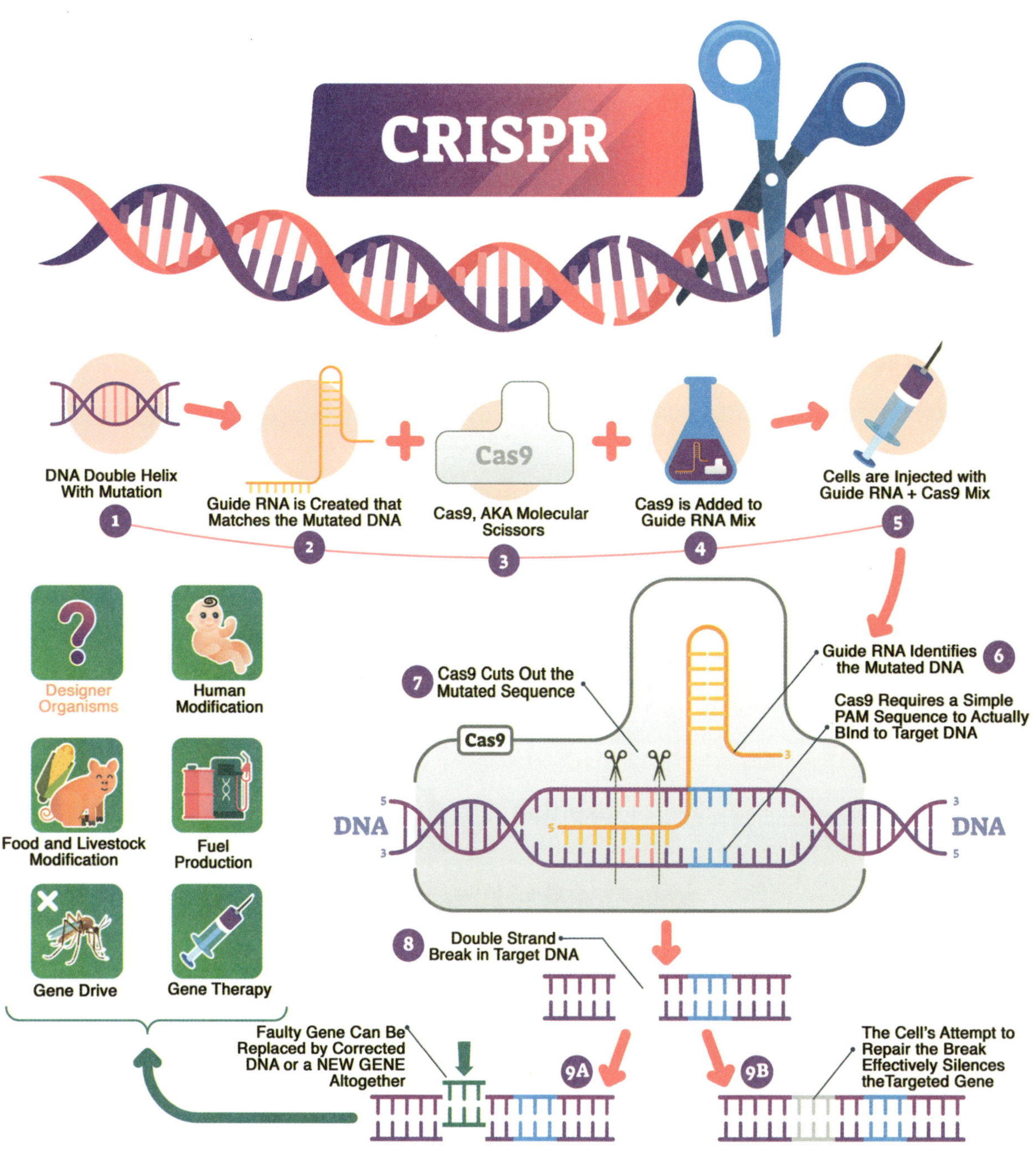

Fig. 9-2.

1. Its potential for cure is greater with diseases that are single gene mutations, rather than those that involve multiple genes and/or environmental causes.
2. There is the problem of how to get the CRISPR complex to the desired place in the body. How do you reach every cell in the body?
3. CRISPR does not always do an exact excision and cutting, so there is the risk of causing more harm than good, especially if it causes a framework mutation, or results in a worse disease, such as cancer.
4. When CRISPR is injected into the body, there is the problem of keeping it stable enough so that it survives long enough to do what it is supposed to do.
5. CRISPR raises ethics questions about manipulating embryos. In fact, the entire idea of using any gene modifying technique has generated strong views about how it should be used if it should be used

at all. Should gene therapy be confined to serious genetic diseases or include less serious ones, for which the (present) risk of potential CRISPR side effects may weigh against the procedure? Will CRISPR be used to enhance the human genome in ways that are not necessary from a medical standpoint to attempt to enhance athletic ability or intelligence, or create a "superhuman"? Will there be inequality in who will receive the enhancement, which may be expensive? Is it wrong to play with Nature in this way?

6. The protein that a gene codes for has to be carefully managed as to amount, timing of production, and cell type in which it is active, to maintain homeostasis. This is challenging.

Here are some ways around these problems:

a. Apply the CRISPR technique to the newly formed embryo and then implant the embryo in the uterus. In that way, every cell that originates from the embryo will contain the favorable mutation. And the disease will no longer be propagated to future generations. And the implant will not be subject to graft rejection. This approach will not help, though, in treating someone who is already full grown. Also, the potential for damage to the embryo is presently significant, and there is much resistance to manipulating embryos in this way.
b. Work with diseases that affect specific organs (e.g. *Huntington disease* in the brain, *cystic fibrosis* in the lungs, *liver cancer*) so that CRISPR is directed exactly where it has to go. In some cases direct injection of CRISPR into the organs that are affected may be effective. *Duchenne muscular dystrophy*, for instance, is an X-linked recessive condition that results from a single gene mutation that prevents the production of the protein *dystrophin*, needed for muscle function. This leads to progressive muscle weakness. Correcting this mutation with CRISPR by direct muscle injection or delivery via the blood stream offers the possibility of improvement or cure.

In the case of *Huntington disease*, a degenerative disease of the brain resulting from an abnormal misfolded protein produced by the *Huntingtin gene*, administration of drugs via the spinal fluid is an option. A trial currently being explored is the spinal fluid administration of a microRNA that blocks the formation of Huntingtin protein by the neural cell's RNA.

Not every genetic disease requires correction in every cell to achieve an improvement. Although the mutation may exist in all the body cells, it is only certain cells and organs in which the gene is expressed. Correcting only a small fraction of cells in the right place may be beneficial.

When a disease is one of blood cells (e.g. *sickle cell anemia, beta-thalassemia, severe combined immunodeficiency*), there is the possibility that the patient can provide his/her own blood cells or bone marrow, and CRISPR can be applied to the cells, which are then infused back with the correction. This method also has the advantage of avoiding an autoimmune response against the donor cells, since they originate in the same person, who is both donor and recipient.

c. Use a suitable CRISPR-carrying vector, like a virus that can reach and penetrate cells, but has been modified to remove its disease-causing qualities. *Adenoviruses* do not integrate into the genome, thereby reducing the possibility of damaging the DNA, but may only act for a short time (especially in dividing cells), requiring repeated applications, which may generate an immune response. *Adeno-associated viruses* do not cause a significant immune response and generally do not incorporate into the genome, but are less able to infect target cells than are adenoviruses. *Lentiviruses*, which can reach and incorporate into the nuclear DNA (and are the ones that transmit HIV), can be modified to become effective vehicles for CRISPR without their harmful effect. However, in some cases, there is concern that the treatment may result in the reactivation of the virus messenger, or induce the development of cancer or other genetic abnormality by altering the DNA, or just not work at all.
d. Use an *artificial vector* (e.g. a *plasmid*, a naked circular chromosome vector that is self-replicating and capable of carrying the CRISPR vehicle) that can easily penetrate cells. *Artificial lipid nanoparticles* can also be used; in addition to being effective vectors, they can degrade quickly after delivering CRISPR, not hanging around to cause unintended problems.
e. *CRISPR prime editing* is a newer variation of CRISPR in current development. It does not require cutting through the double strand of DNA before making the nucleotide substitution, which is subject to significant errors. Instead, it nicks only one of the two DNA strands, which is enough to allow the nucleotide or nucleotide sequence substitution.
f. The cutting ability of CRISPR/Cas9 can be disabled, and an activator or repressor transcription factor, or an epigenetic modifier can be attached to CRISPR to alter gene expression rather than cutting the DNA chain.
g. Instead of using CRISPR to act on DNA,

significant research is underway to affect protein synthesis by interfering with mRNAs, e.g. the introduction of select microRNAs to disable particular mRNAs. MicroRNAs bind to and silence corresponding mRNAs, thereby inhibiting the mRNA's translation and protein production.

Cloned stem cells are *multipotent* and can potentially replace diseased tissues or organs. If the cells/tissues that are cloned belong to the patient, this eliminates the problem of graft rejection. Most cells in the course of development normally change from a state of multipotency to a state of *pluripotency*, where they can change to only selected kinds of tissues, to a state of *determination*, where they can only form one particular tissue. Techniques to convert cells that are determined into multipotent cells that can be used as stem cells are a promising avenue for gene therapy.

Figure 9-3 lists some diseases for which gene therapy trials have been initiated. Updates on current progress may be found at *https://clinicaltrials.gov.*

FIGURE 9-3. SOME DISEASES UNDERGOING GENE THERAPY TRIALS
Achondroplasia
Adenosine deaminase severe combined immune deficiency
AIDS/HIV
Alzheimer disease
Cancers (colorectal, cervical, ovarian, melanoma, glioblastoma, hepatocellular, prostate, pancreatic, lung, breast, renal)
Cerebral adrenoleukodystrophy
Choroideremia
Chronic granulomatous diseases
Congenital hearing loss
Cystic fibrosis
Diabetes
Fragile X
Heart failure
Hemophilia
Huntington disease
Hypercholesterolemia
Inborn errors of metabolism
Infertility
Leber hereditary optic neuropathy
Metachromatic leukodystrophy
Muscular dystrophy
Mucopolysaccharidosis Sanfilippo type A
Myotubular myopathy
Parkinson disease
Sickle cell disease
Spinal muscular atrophy
Tay-Sachs disease
Thalassemia
Wiskott-Aldrich syndrome

Treatment of Cancer

For many years the focus of cancer treatment has involved chemotherapy and/or radiotherapy, the idea being that the drug or radiation would damage malignant cells to a greater extent than normal body cells, by killing cells that grow and divide quickly, like cancer cells. Normal cells also have protective repair mechanisms that make them less vulnerable to chemotherapy and radiation damage. While frequently curative, such treatments have often been marked by significant side effects in the process, since the treatments also damage normal cells to some degree.

Targeted therapy is a form of chemotherapy that is more specific, focusing on differences between normal cells and the specific type of cancer cell being treated. The targeting drug may block certain chemical signals that tell the particular cancer cell to grow and divide; interact with specific proteins in the cancer cell; inhibit the growth of new blood vessels so that cancer cells are less able to grow; carry select toxins to the cancer cells that will kill them but not normal cells; or stimulate the immune system to better attack cancer cells. An important aim is to try to develop drugs that are specific inhibitors, rather than pan inhibitors, since we would expect pan inhibitors to have more side effects.

One of the main problems with some targeted therapies, apart from their own potential toxic effects on normal cells, occurs when the patient develops resistance to the treatment.

More recently, it has come to attention that there may be more epigenetic abnormalities in cancer cells than mutations. Epigenetic (environmental) influences include diet, obesity, drugs, tobacco and alcohol intake, pollutants, and other chemicals that have a lasting effect on DNA expression without changing the DNA nucleotide sequence. Since epigenetic abnormalities may be reversible by removing the environmental influence, this provides a more encouraging approach to treatment.

Differentiation therapy is based on the idea that malignant cells act like immature cells that do not differentiate but, as in the embryo, continue to divide. Differentiation therapy uses pharmacologic agents to change cancer cells into a normal mature differentiated state, with less toxicity than standard treatment protocols. For instance, *all-trans-retinoic acid (ATRA)* has been used successfully in treating acute *promyelocytic leukemia (APL)*, a condition in which there is a defect in the retinoic acid receptor alpha gene, resulting in a defect in myeloid cell differentiation.

Immunotherapy

The body normally prevents cancer by two kinds of mechanism, one internal to the cell and one external.

Internally, cancer is normally kept under control by genes that enhance tumor suppression and DNA repair, and prevent excessive cell growth and proliferation. *Externally*, cancer is normally kept at bay by the immune system, which detects abnormal proteins (or the absence of normal proteins) on cancer cells and specifically attacks those abnormal cells. Cancer immunotherapy largely aims to boost the immune system so that it is better able to fight the cancer. Part of this boost is the general activation of the immune system, as well as the production of antibodies that can specifically target tumor cells. One method is called *CAR-T (Chimeric Antigen Receptor T-cell therapy)*, which provides immune cells with an artificial gene that allows the cells to target and attack cancer cells. Chemotherapy and immunotherapy can be used in combination.

Cancer treatment today is personalized medicine, since each person has his/her own unique genome, and understanding a person's unique DNA sequence helps taper therapy to meet individual needs.

Gene therapy with CRISPR is a burgeoning and promising area of cancer research that ties in well with immunology. CRISPR has the capacity to disable genes in cancer cells that allow the cells to evade the immune system, or that cause excess cell proliferation. CRISPR can also genetically enhance immune cells, enabling them to mount a more vigorous and specific immune response. Altering the genes helps insure a lasting change, which is more effective than an exogenously administered drug or virus that does not incorporate into DNA and whose effect diminishes with time. Current studies are giving close attention to determining how strong and lasting the therapies will be and how free they will be of adverse side effects.

Pharmacogenomics

Pharmacogenomics (also called *pharmacogenetics*) explores how the person's genome influences the effects of drugs. It is a major area of research with huge therapeutic potential for personalized medicine, where the type of drugs and their dosage can be tapered to fit the patient's needs, depending on the patient's genome. Metabolic differences in one person can make a drug more active or less active than in another person. If an enzyme acts too quickly, the drug may be inactivated too fast, and the drug will not be effective. If the enzyme acts too slowly to degrade the drug, and the drug hangs around, the drug may have too much effect, even build to toxic levels. Examples of individual differences to specific drugs:

- People who are *fast acetylators* need higher or more frequent doses of acetylated drugs, e.g. *isoniazid* for tuberculosis.
- People who are *slow acetylators*: Adverse effects may occur from acetylated drugs, e.g. isoniazid (peripheral neuritis), hydralazine or procainamide (lupus).
- *CYP2C gene polymorphism* (over 50 different types): There may be reduced action of *clopidogrel*, with increased risk of thrombosis due to a decrease in antiplatelet action.
- *G6PD deficiency*: Risk of hemolytic anemia with oxidant drugs, e.g. chloroquine, primaquine antimalarials.
- *HLA-B*1502 gene*: *Stevens-Johnson syndrome* - Severe dermatologic reaction to carbamazepine (anticonvulsant/mood stabilizer).
- *Plasma pseudocholinesterase deficiency*: Succinylcholine inactivation is decreased, leading to prolonged paralysis of respiration during general anesthesia.
- *HLA-B*5701 gene*: Hypersensitivity (rash, fatigue, diarrhea) to antiviral drug abacavir, used for HIV infection.
- *Thiopurine S-methytransferase* gene mutation: Azothioprine (drug used for rheumatoid arthritis) cannot be properly converted and can build up in bone marrow.

The rates of absorption, excretion, metabolism and effects of a drug commonly differ among patients and are affected by variables such as the patient's sex, age, weight, use of alcohol, smoking, exercise, and genome. It would be nice to have a drug individualized to the patient, rather than one size fits all, or relying on trial and error to find the best drug for the patient.

Pharmacogenomics is not an easy field. It can be hard to interpret pharmacogenomic studies since there may be multiple genes that influence a drug's effect. And you have to consider the interactions of the selected drug with other drugs and environmental factors, for which adequate evidence may not be available. There are also issues of the expense of the evaluation and treatment, whether insurance companies will pay for it, and honesty and reliability of certain companies trying to sell pharmacogenomics to the public.

10

Homeobox Genes

Many thousands of biochemical reactions occur in an individual cell at any given time. Each cell seems to know its proper place and function in the body. But what coordinates all this activity? Is there a master control that directs the creation of the body pattern in embryonic development, so that each cell knows its proper body position during a lifetime of growth, regeneration, and continual metabolic changes? Is the orderly mechanism like a complex government, with a leader who directs various societal divisions that control their underlying organizations, which in turn feed back information to the leaders above them? Or is it more like a complex ecosystem, like a forest, with no particular leader, but with numerous components that interact in a way that maintains homeostasis? The forest responds to the environment in a balanced manner by having undergone evolution over millions of years in a way that promotes survival.

What is it that directs body development to such a refined degree that a child looks like the parent?

There does not appear to be a single leader that oversees all the cells and their patterning, like a brain overseeing the activities of the nervous system. All the instructions for regulating body function reside in the genes, and every nucleated cell among the trillions of cells in the body has its own complete set of genes and does its own thing. The difference between cells lies in the turning on or off of the cell's specific genes in a way that uniquely provides for the needs of that individual cell. This individuality is largely due to the production by the genes of thousands of different transcription factors, proteins whose function is that of interacting with other genes in the cell and regulating their expression. Different transcription factors function differently in different cells.

But what makes each individual cell so unique as to manifest its own special quantity and quality of transcription factors?

Part of the answer lies in **homeobox genes**. These genes have been conserved through evolution and are strikingly similar in very different organisms, from bacteria and fruit flies to people. Most homeobox gene proteins are transcription factors, which regulate the expression of other genes by attaching to DNA via a region in the homeobox transcription factor protein called a **homeodomain**. A single homeobox transcription factor may regulate a multitude of genes, increasing (turning "on") or decreasing (turning "off") gene expression of other transcription factors. Conversely, a number of different transcription factors can affect the same gene. The cell is a hub of incredibly complex activity, much like the people in a large city interacting with one another. Numerous chemical interactions go on in a cell every second. Their interactions resemble a massive web of interactivity, rather than a simple linear set of biochemical reactions. Since many of these reactions are "housekeeping" functions present in all cells, a mutation that affects one part of the web can not only disrupt one cell's function, but disrupt functions throughout the body where the gene is expressed. Sometimes a gene is expressed in only certain parts of the body. For example, in *Holt-Oram syndrome,* there are abnormalities in the development of the upper limb bones and heart. There may be absent arm, wrist and thumb bones, as well as cardiac atrial or ventricular septal defects and cardiac arrhythmias. These separate anomalies occur because the genes affecting the upper extremities and heart development are expressed particularly in those body areas as opposed to other areas.

HOX genes (of which there are some 39 in humans, located in four clusters on different chromosomes) are a subgroup of homeobox genes (of which there are over 200 in humans). Both kinds of genes produce transcription factors. While the HOX genes are more focused on the specification of the order of major body segments along the anterior-posterior (head-to-tail) axis in embryonic development, the homeobox genes as a whole have a broader function in regulating a wider variety of body functions during and after embryonic development.

It is not as if the HOX genes have a brain that is aware of and directs all the activities in its cells. HOX genes are more of a coordinating unit that generally directs the position of the segments of the body in embryogenesis, but does not specify the actual composition of those segments. Other genes do that. I.e., while both the fruit fly and other bilateral animals, including humans, all have similar HOX genes, which are important in directing the relative position of the various body segments, the body segments themselves are quite different in the fruit fly and in a person, and depend on other genes for their unique characteristics. In the fruit fly, for instance, mutations to HOX genes may result in a disturbance in development of body segments with such abnormalities as a leg growing in place of an antenna. In humans, a HOX mutation may result in the widespread distribution of congenital defects in a given genetic disease (see the examples at the end of this chapter).

The so-called "brains" of the whole homeobox transcription operation lie hidden in the organization of the interactions of transcription factors in the web of intercellular and extracellular activity. This organization is more like the complex forest ecosystem, which maintains its balance through the interactions of its components with themselves and the surrounding environment, without a central brain.

Since homeobox genes and the transcription factors they produce have such broad influences, defects in those genes in people, as one would expect, may result in significant dysfunction. Some abnormalities may be more localized to the specific body areas where the genes are expressed, such as the limbs (HOX gene mutations typically cause congenital malformation of the limbs), axial skeleton, eyes, head, face, teeth, or GI, urogenital, or nervous systems. The positive clinical take-home point is that a medical syndrome can include multiple areas of bodily dysfunction, but arise from a single mutation. If that mutation can be corrected, it will affect all the areas of dysfunction. Of course, it may be too late to make the correction once the child is already born with a congenital malformation, but there are many diseases in which gene therapy may become an option.

The kinds of functions that homeobox genes help regulate include *cell growth, division, differentation, migration, shape, positioning,* and *cell death.* Many diseases, including many cancers, are associated with homeobox mutations or abnormal expression.

How did such a complex, perfectly balanced system arise? Is there a creator or an unknown and unseen new type of force in Nature? Or is it a natural development through millions of years of evolution? I don't know (but lean toward a naturalistic explanation).

The blueprint for turning gene expression "on" or "off" is incredibly complex. There are thousands of transcription factors and thousands of enhancer and repressor sites along the DNA, where transcription factors bind in different combinations. How is everything in the body controlled so accurately?

There are two kinds of codes associated with DNA. One is the classic genetic code of the base sequence of DNA, which results in particular proteins. Each DNA nucleotide triple base sequence is like a letter, and each protein resulting from stringing the bases into a protein chain in the ribosomes is like a word.

However, there is a *second code,* largely unknown, which corresponds to the book constructed from the words, the distinct phenotype that arises from the genome. The code for creation of the book, in a way, resembles a computer program, with "on" and "off" circuit switches (**Figure 10-1**; compare with **Figure 3-11**).

Figure 10-1. Computer circuits for "and" and "or." The "and" switches are those switches that are in series. If *both* series switches close, an "and" arrangement emerges, and the light in this case turns "on"; the

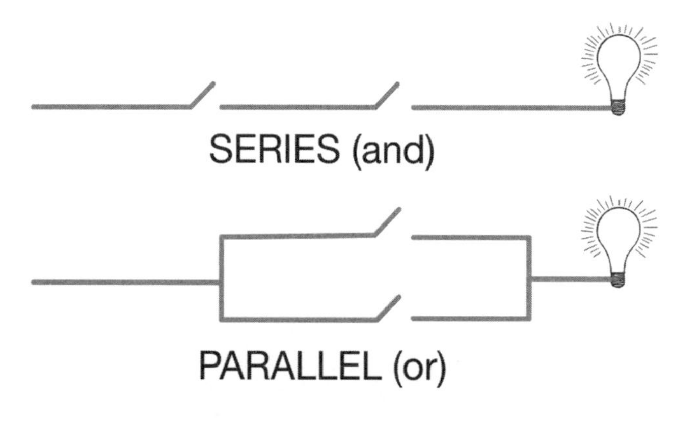

Fig. 10-1.

message can get through. Otherwise, the light is "off." "Or" switches are parallel. If *either* of the parallel switches closes, that suffices to send the message through. Computer programs are constructed largely with a combination of "and/or" circuits. The greater the complexity of the circuit structure, the greater the complexity of the computer program (consider the complex code and actions in a modern computer game). In the cell, the "on/off" circuits are the responses of DNA to the combined actions of transcription factor activators/repressors on DNA promoters/enhancers/silencers, which determine whether the gene is expressed or not (see **Figure 3-11**). Some gene expression may be more of a series (*and*) arrangement, where a number of different transcription factors need to act at once to allow a gene's expression. In other cases, gene expression may occur with the action of either one transcription factor or another (a parallel *or* arrangement). The collective action of multiple transcription factors helps determine the unique expression of each cell in the body. The code is written in the DNA of each cell, and unfolds in a symphony of serial and parallel processing to produce the phenotype in time and space.

How does gene expression come to be selectively localized to particular body regions?

The genetic "computer" program involves more than just the expression of particular proteins, but their expression in *time*. After the division of the zygote, each daughter cell from the division acquires some difference from the parent cell in the form of a different pattern of DNA expression, depending not only on the kind of protein that is produced but on its quantity and timing with other cell events. As the body grows, it develops enormous differences in the gene expression in its different cells. Some of the genes become expressed at different times and places than genes in other cells, if they are expressed at all. Each cell has its own *positional information* in the body. Since different body areas have different gene expressions, a mutation to one gene may affect one body area, where the gene is expressed, to the exclusion of others.

The timing of the code in most computer programs is written in serial fashion. That is, the computer code is a series of instructions as to what should happen when the previous instruction is carried out. Commonly, a bug in the program can interrupt this serial chain, and the entire program stops functioning, like all the lights may go out when one bulb goes out in a serial chain of lights. This is different from *parallel* programming, in which a number of separate instructions act independently of one another. Think of leaves on a tree. Each leaf follows its own set of instructions, and the tree does not die because one leaf is removed. The trillions of cells in the body are like leaves, following their independent courses. Virtually all of them contain a full complement of DNA. At the same time, they communicate with nearby cells, whether through direct cell-to-cell contact or diffusion of proteins (some with strange but actual names like *sonic hedgehog*, *wingless*, *frizzled*, and *noggin*) to signal changes in nearby cells via a chemical gradient (**paracrine signaling**). The concentrations of chemicals in the gradients can influence how the genes of the nearby cells are expressed. Cells may also be influenced by circulating hormones, which act across a significant distance. In addition, cells form organs, which interact with one another. The kidneys, lungs, heart, blood circulation, and nervous system, continuously interact to maintain order in the body. Thus, trillions of body cells, each with thousands of interactions happening inside each one of them, all interact in one way or another with one another, a massive symphony of serial and parallel processing, to create the story book, the phenotype. Numerous researchers have conducted painstaking research in defining the words and some of the story of the chemical interactions, but an understanding of the overall book is still far off. There are 26 letters and a quarter of a million words in the English language, but billions of stories. The more we understand, the greater the feasibility of treating genetic diseases.

Biochemical interactions in the cell and body are a huge interlocking web of associations. Mutations that interrupt different areas of this web can affect similar functions ("All roads lead to Rome"). Moreover, a single mutation can affect multiple functions. Attempting to fix a homeobox mutation with gene therapy holds great promise for correcting a number of abnormalities at once, but at this time presents a challenge in finding and correcting the gene early enough in the course of the disease, without significant side effects.

Examples of homeobox genes and their function:

ADNP: Brain development (may involve alteration in number of neurons or development of normal connections)
Mutation: *Autism spectrum disorder, ADNP (Activity-Dependent Neuroprotective Protein) syndrome*; cancer through association with cell growth and differentiation.

ALX1: Head and face, eyes, nose, mouth. Controls activity of genes that regulate cell growth, division, and movement, ensuring that cells grow and stop growing at specific times and that they are positioned correctly during development.
Mutation: frontonasal dysplasia; associated with lung cancer.

ALX4: Skull, skin
Mutation: Enlarged parietal foramina, frontonasal dysplasia (face, nose, clefts, hair loss,

genital abnormalities in males), *Potocki-Shaffer syndrome* (same as previous plus intellectual disability, defects in heart, kidneys, urinary tract); as a tumor suppressor gene, it is associated with breast, lung, and liver cancer.

ARX: Brain, pancreas, testes, and muscles
Mutation: *Early infantile epileptic encephalopathy 1, Partington syndrome, X-linked lissencephaly with abnormal genitalia.*

CRX: Cone-rod homeobox protein binds to specific regions of DNA and helps control the activity of particular genes.
Mutation: *Cone-rod dystrophy, Leber congenital amaurosis, retinitis pigmentosa.* Associated with retinoblastoma and pineal tumors.

HESX1: Brain, pituitary, optic nerve development
Mutation: *Combined pituitary hormone deficiency, septo-optic dysplasia,* genital abnormalities, vision/facial abnormality, supernumerary fingers.

HOXA13: Limbs, urinary tract, reproductive system
Mutation: *Hand-foot-genital syndrome,* cancers (early onset prostate cancer).

HOXB13: Skin maintenance, tumor suppressor
Mutation: Increased risk of prostate cancer.

LHX1: Brain, female reproductive system
Mutation: Renal cysts and diabetes syndrome, *Kitansky-Kuster-Hauser syndrome* (underdevelopment of vagina and uterus), abnormalities of kidneys, urinary tract, reproductive system, *maturity onset diabetes of the young type 5,* intellectual disability, behavioral/psychiatric disorders. Promotes migration and invasion of lung cancer cells.

LMX1B: Limbs, kidneys, eyes
Mutation: *Nail-patella syndrome* (abnormalities of nails, knees, elbows, pelvis). High expression promotes migration of cancer cells.

MEOX1: Directs formation of body structures in developing embryo
Mutation: Spinal fusion, incorrect positioning of vertebral (scoliosis) and development of many other parts of body (hearing, eye, cleft palate, genitourinary, heart, lung), limb length, brain; high expression found in lung cancer.

NSX1: Teeth, mouth, fingernails and toenails
Mutation: Cleft lip/palate, *Wolf-Hirschhorn syndrome* (facial abnormalities, delayed growth, intellectual disability, seizures).

MSX2: Bone growth, skull
Mutation: Enlarged parietal foramina, craniosynostosis.

NKX2-1: Brain, lungs (surfactant), thyroid gland (thyroid hormones)
Mutation: *Brain-lung-thyroid syndrome* (chorea); high expression in certain lung cancers.

OTX2: Ocular development, optic nerves, brain development, pituitary
Mutation: *Septo-optic dysplasia,* coloboma, combined pituitary hormone deficiency, microphthalmia; associated with medulloblastoma cerebellar cancer.

PAX3: Acts on neural crest cells, face and skull bones, muscles
Mutation: *Craniofacial-deafness-hand syndrome, Waardenburg syndrome* (hearing loss, changes in pigmentation of hair, skin, eyes), cancers (rhabdomyosarcoma).

PAX6: Eyes, brain, spinal cord, pancreas, olfactory bulb
Mutation: Aniridia (absent iris), *Peters anomaly* (cloudy cornea, coloboma, microphthalmia), *WAGR syndrome* (Wilms tumor, aniridia, genitourinary anomalies, intellectual disability).

PHOX2A: Neuron differentiation, autonomic nervous system, CN III and IV
Mutation: Congenital fibrosis of the extraocular muscles; neuroblastoma.

PHOX2B: Neural crest (autonomic nervous system)
Mutation: *Congenital central hypoventilation syndrome* (some have *Hirschsprung disease*), neuroblastoma.

PITX1: Lower limbs. The PITX1 protein is found primarily in the developing legs and feet, pituitary gland, branchial arch structures (roof of mouth, jaw, parts of inner ear).
Mutation: *Liebenberg syndrome* (short fingers, elbow and wrist contractures, upper extremities develop more like the foot), clubfoot, shortened tibia, polydactyly. A tumor-suppressor gene, downregulated in a number of cancers, including gastric, esophageal, colorectal, and lung.

PITX2: Anterior segment of eye (cornea); may act in embryonic or adult development.
Mutation: *Axenfeld-Rieger syndrome* (also has distinctive facies, tooth abnormalities, other

body areas affected), *Peters anomaly.* Overexpression in colorectal cancer.

POU3F4: Determines cell types in brain and spinal cord in early development, middle and inner ear
Mutation: Nonsyndromic hearing loss, sensitivity to cold, PLAID syndrome—antibody deficiency, cold-induced urtricaria, immune dysfunction, frequent colds, pneumonia, autoimmune thyroiditis, vitiligo (pigment loss in skin), autoantibodies.

PROP1: Pituitary gland (specialization of cell types)
Mutation: Combined pituitary hormone deficiency; mutation associated with craniopharyngiomas of pituitary gland.

SHOX: Skeleton
Mutation: *Langer mesomelic dysplasia* (short stature, shortening of long bones in arms and legs, abnormality of wrist and forearm bones), *Leri-Weill dyschondrosteosis* (similar to *Langer*), *Mayer-Rokitansky-Kuster-Hauser syndrome*, *Turner syndrome*; expressed in certain tumors.

SIX1: 2nd branchial arch, thymus, skeletal muscle
Mutation: *Branchiootorean/branchiootic syndrome* (malformed ears, neck; congenital anomalies of kidney and urinary tract); aberrant expression in multiple cancers.

SIX3: Helps establish the right and left halves (hemispheres) of the forebrain, lens of eye, retina.
Mutation: *Nonsyndromic holoprosencephaly* (brain fails to divide into two hemispheres), coloboma; loss of SIX3 expression correlated with tumor progression.

SIX5: 2nd branchial arch, neck, ears, kidneys, brain, heart, eyes, skeletal muscles
Mutation: *Branchiootorenal/branchiootic syndrome*, congenital anomalies of kidney and urinary tract.

TGIF1: Blocks signals in TGF-beta pathway
Mutation: *Nonsyndromic holoprosencephaly* (Brain doesn't divide into two halves. Sometimes cyclopia, microcephaly, hydrocephalus, cleft palate, single front tooth, microphthalmia, developmental delay, intellectual disability, pituitary malfunction). Overexpression associated with certain cancers.

ZEB2: Neural crest, digestive tract, skeletal muscles, kidneys
Mutation: *Mowat-Wilson syndrome* (abnormal development of neural crest-derived structures, e.g. nervous system, facial, digestive tract, coloboma). The role of the ZEB2 protein in the development of nerves that control the digestive tract may help explain why many people with this condition also have *Hirschsprung disease*, an intestinal disorder that causes severe constipation, intestinal blockage, and enlargement of the colon. ZEB2 expression promotes gastric cancer cell migration and metastasis.

Appendix. Diseases

This appendix provides further information on the medical conditions mentioned in this book, including inheritance, major features, and treatment. AD = autosomal dominant; AR = autosomal recessive; XD = X-linked dominant; XR = X-linked recessive; Rx = treatment.

3-beta-hydroxysteroid dehydrogenase deficiency (AR) Sodium loss in urine; genital abnormalities. **Rx**: Replacement of glucocorticoids, mineralocorticoids, and sex steroids.

3-Methylcrotonyl-CoA carboxylase deficiency (AR) Cannot break down leucine-containing proteins. Feeding difficulty, lethargy, hypotonia, delayed development, seizures; some people asymptomatic. **Rx**: Low-protein diet.

11-beta hydroxylase deficiency (AR) Excess androgen production, premature sexual development, hypertension. **Rx**: Antihypertensives; glucocorticoid replacement.

17-alpha hydroxylase deficiency (AR) Hypertension, hypokalemia, abnormal sexual development. **Rx**: Blood pressure control, glucocorticoid replacement.

17,20 lyase deficiency (AR) Impaired production of androgen and estrogen. Underdeveloped secondary sexual characteristics. **Rx**: Hormone replacement with androgens and estrogens, surgical correction of undescended testes.

18-dehydrogenase and **18-hydroxylase deficiency** (AR) Salt-losing through decreased aldosterone; dehydration, vomiting, stunted growth. **Rx**: glucocorticosteroids, mineralocorticosteroids, dietary salt.

3-Methylcrotonyl-CoA carboxylase deficiency (AR) Cannot break down leucine-containing proteins; infants have feeding difficulties, lethargy, hypotonia, seizures. **Rx**: low-protein diet.

21-hydroxylase deficiency (AR) Salt wasting and virilizing forms. **Rx**: Hormones and/or steroid replacement; surgery for ambiguous genitalia in females; psychotherapy.

Abetalipoproteinemia (AR) Cannot absorb dietary fats; failure to thrive, diarrhea, abnormal red blood cells, neurological problems. **Rx**: Vitamin E and vitamin supplements with fat-soluble vitamins, low-fat diet; neurology consultation in some.

Achondroplasia (AD) Short-limbed dwarfism, large head, normal intelligence, cartilage in long bones cannot change to bone. **Rx**: Growth hormone, surgery for spinal lordosis/scoliosis.

Adenosine deaminase deficiency (AR) Pneumonia, chronic diarrhea, skin rashes, slow growth, developmental delay. **Rx**: Treatment of infections, IV immunoglobulins, bone marrow transplant, enzyme replacement, gene therapy.

ADNP syndrome (Activity-Dependent Neuroprotective Protein) (AD) Autism spectrum disorder, intellectual deficit; can affect multiple systems in body, happy demeanor. **Rx**: Symptomatic, treatment for specific symptoms (eyeglasses/surgery for eye problems, physical/speech/occupational therapy, seizure medication, sleep/behavioral medication) Trial of CP201/NAP neuroprotective protein.

Albinism (tyrosinase deficiency) (AR) Color loss in hair, skin, eyes; light sensitivity. **Rx**: Skin/eye care.

Adrenal gland cancer (AD or AR) **Rx**: Surgery, radiation, chemotherapy, hormone modification.

Aicardi syndrome (XD) Almost all in females. Dysgenesis of the corpus callosum of brain; seizures, chorioretinal defects, microcephaly, microphthalmia, coloboma of optic nerve, small hands, facial features; scoliosis in some. **Rx**: Antiepileptic medication, counseling.

Aicardi-Goutieres syndrome (AR) Encephalopathy; nervous system inflammation, muscle spasticity, autoimmune disorders, itchy skin lesions. **Rx**: Physical/speech therapy, correction of scoliosis, diabetes insipidus, underactive thyroid, glaucoma, cardiomyopathy, pulmonary hypertension, and bleeding tendency.

AIDS/HIV CCR5-delta gene mutation (AR) provides *resistance* to AIDS immune system dysfunction and its susceptibility to infections.

Albright hereditary osteodystrophy (AD) Short stature, brachydactyly, subcutaneous ossification, resistance to parathyroid hormone. **Rx**: Calcium/vitamin D supplements, low-phosphorous diet, phosphate-lowering medication.

Aldolase deficiency (AR) Fructose intolerance, seizures, anemia, sleepiness, infant feeding problems, kidney dysfunction, growth retardation, myopathy, exercise intolerance. **Rx**: Supportive; red cell blood transfusions/ splenectomy as needed for anemia. Reduce sucrose, fructose, sorbital from diet.

Alkaptonuria (defect in homogentisate dioxygenase enzyme) (AR) Thick, dark ear cartilage (ochronosis), dark sclera, blue skin discoloration, black earwax, dark urine, kidney/prostate stones, fused vertebrae, decreased joint mobility. **Rx**: Low-protein diet, vitamin C, anti-inflammatory medication for joint pain, physical/occupational therapy, joint surgery; nitisinone to reduce homogentisic acid accumulation by inhibiting homogentisic dioxygenase.

Alpha-1 antitrypsin deficiency (AR) Alpha-1 antitrypsin, normally produced in the liver, protects the lungs against damage from elastase. Low levels of alpha-1 antitrypsin result in emphysema; also liver damage (cirrhosis) from misfolded alpha-1 antitrypsin protein; risk of hepatocellular carcinoma. **Rx**: Bronchodilators, intravenous alpha-1 antitrypsin; lung transplant in some.

Alpha-thalassemia myelodysplasia syndrome (XR) Mostly in males. Anemia, weakness, shortness of breath, susceptibility to bleeding/infections. **Rx**: Blood transfusion, erythropoietin stimulation, antibiotics if needed.

Alport syndrome (XD) Kidney disease, eye abnormalities, hearing loss. **Rx**: Symptomatic, blood pressure lowering medication.

Alstrom syndrome (AR) Progressive vision/hearing loss, cardiomyopathy, obesity Type II diabetes, short stature, liver/kidney/bladder/pulmonary problems, acanthosis nigricans (thick dark, velvety skin creases). **Rx**: Eyeglasses, hearing aids, diabetes/heart/kidney medication, kidney/liver transplant in some.

Amyotrophic lateral sclerosis (AD or AR, only 5-10% inherited) Muscle weakness, atrophy. **Rx**: Reduce motor neuron damage with the drugs riluzole (decreases the body levels of glutamate) and edaravone (an antioxidant).

Alzheimer disease (AD in early onset disease) Dementia. **Rx**: Cholinesterase inhibitors; drugs to decrease beta-amyloid in brain; exercise, nutrition, mental/social engagement; gene therapy.

Amyloidosis (AD) Adult-onset weakness, urinary problems, weight loss, dizziness, dry eyes, difficulty swallowing, easy bruising, irregular heartbeat, neuropathy. **Rx**: Chemotherapy to inhibit production of abnormal TTR protein (which deposits in tissues); corticosteroids.

Anderson disease (deficiency of branching enzyme, Type IV GSD) (AR) Muscle weakness/cramps, hypoglycemic seizures, cardiomegaly, hepatosplenomegaly, liver cirrhosis. **Rx**: Liver transplant in some.

Anemia, congenital hemolytic (AR) Jaundice, enlarged spleen, liver, **Rx**: Blood transfusion, plasmapheresis, removal of spleen, stem cell transplant.

Anemia, congenital hypoplastic (AR) Anemia, congenital anomalies, increased risk of leukemia, sarcoma. **Rx**: Blood transfusion, corticosteroids, bone marrow/stem cell transplant; monitor for leukemia/cancer.

Angelman syndrome (AD, usually not inherited) Delayed development, happy demeanor. **Rx**: Antiseizure medication, physical/communication/behavior therapy.

Aniridia (AD, new mutations) Defect in iris of eye with misshapen pupils. **Rx**: Surgery for glaucoma/cataracts, contact lenses, monitor for Wilms tumor.

Apert syndrome (AD) Skeletal abnormalities, especially skull; vision problems; syndactyly; cognitive dysfunction in some. **Rx**: Surgery for physical deformity; eye drops for dry eyes; CPAP breathing machine at night for some; antibiotics, tracheostomy in some; ear tubes for recurrent ear infections.

Aplastic anemia (The hereditary form is usually associated with other diseases (e.g. *Fanconi anemia*, *Shwachman-Diamond syndrome*, *dyskeratosis congenita* and *Diamond-Blackfan anemia*) Fatigue, shortness of breath, dizziness, prolonged bleeding from cuts, nosebleeds, and gums, pallor. Low counts of red and white blood cells and platelets. **Rx:** Blood transfusion, blood/bone marrow transplant, immune suppressives.

APRT (adenine p-ribosyl transferase) deficiency (AR) Disorder of purine metabolism, kidney stones. **Rx**: Allupurinol, high fluid intake, low-purine diet to reduce kidney stones.

Argininosuccinic aciduria (AR) Ammonia accumulation in infant, lethargy, seizures, developmental delay, liver damage, skin lesions, brittle hair. **Rx**: High-calorie, reduced protein diet, supplementary arginine; sodium benzoate/ sodium phenylacetate to reduce ammonia levels; hemodialysis in some.

Asphyxiating thoracic dysplasia (AR) Growth disorder of multiple bones, short stature, extra fingers/toes, breathing difficulty, kidney malfunction. **Rx**: Manage respiratory infections, monitor renal/hepatic function, corrective bone surgery.

Asthma (AD or multifactorial if inherited) Breathing difficulty, inflammation of airways. **Rx**: Bronchodilators, corticosteroids.

Ataxia telangiectasia (AR) Ataxia, enlarged skin blood vessels, weak immune system, leukemia/lymphoma risk. **Rx**: Antibiotics, monitor for cancer; gene therapy.

Athabascan brainstem dysgenesis syndrome (AR) Horizontal gaze palsy, sensorineural deafness, central hypoventilation, developmental delay. **Rx**: Oxygen, mechanical ventilation.

Autism spectrum disorder (Multifactorial, sometimes X-linked) Impaired social skills, repetitive behaviors, psychiatric problems, eating disorders, may be a feature of other syndromes. **Rx**: Varies with patient. Home/school behavioral intervention; treatment as needed for GI/feeding problems, sleep disturbances, ADH disorder, anxiety/depression/obsessive compulsive disorders, epilepsy.

Autoimmune polyglandular syndrome (AR) Thyroid/adrenal dysfunction, Type I diabetes mellitus. **Rx**: Hormonal replacement.

Autoimmunity (AD) **Rx**: Anti-inflammatory drugs.

Axenfeld-Rieger syndrome (AD) Abnormal iris, glaucoma, facial/dental abnormalities, may be hypospadias (urethra opening is on underside of penis), anal stenosis, pituitary abnormality. **Rx**: Treat glaucoma if present; surgery for facial/dental problems; heart surgery; corrective surgery for hypospadias/anal stenosis in some.

Baller-Gerold syndrome (AS) Prematurely fused skull bones, abnormal head shape, missing fingers, small stature. **Rx**: Surgery for defects.

Bardet-Biedl syndrome (AR) Bone abnormalities of skull, arms and hands; slow growth; small stature; defects of kneecaps; patchy skin rash. **Rx**: Surgery for bone defects.

Basal cell nevus syndrome (AD) Skeletal anomalies, basal cell carcinoma. **Rx**: Surgery.

Beckwith-Wiedemann syndrome (AD) Asymmetric body growth, abdominal hernia, hypoglycemia, Wilms tumor of kidney, hepatoblastoma. **Rx**: Medication for hypoglycemia; surgery for hernia repair, enlarged tongue.

Behcet disease (Sporadic, no clear hereditary history) Inflammatory sores of mouth, genitals, elsewhere (erythema nodosum), joints, brain, uveitis. **Rx**: Immunosuppressive agents.

Bernard-Soulier syndrome (AR) Bleeding disorder with abnormal platelets. **Rx**: Desmopressin (releases clotting factor VIII from platelets), antifibinolytic agents if needed.

Beta-ketothiolase deficiency (AR) Infant can't process isoleucine; lethargy, dehydration, breathing difficulty, seizures. **Rx**: Manage acute crises with IV fluids, glucose, electrolytes, bicarbonate.

Beta-thalassemia (AR) Child develops anemia, failure to thrive, jaundice, enlarged spleen, liver, heart. **Rx**: Blood transfusion, iron chelation therapy, spleen removal, folic acid, gall bladder removal, bone marrow transplant.

Biotinidase deficiency (AR) Infant can't process biotin; develops seizures, hypotonia, breathing, hearing, and vision problems, ataxia; hair loss; skin rashes. **Rx**: Oral biotin supplementation.

Birt-Hogg-Dubé syndrome (AD) Risk of development of benign and malignant tumors of skin and kidney; pneumothorax. **Rx**: Remove skin lesions.

Bloom syndrome (AR) long face, narrow nose, short stature, increased cancer risk, sun sensitivity, patchy skin pigmentation. **Rx**: Sunscreen, antibiotics for infections.

Bone cancer (Increased risk in Li-Fraumeni syndrome, hereditary retinoblastoma, Paget's disease of bone). **Rx**: Surgery, chemotherapy, radiotherapy, cryosurgery, targeted therapy.

Bosley–Salih–Alorainy syndrome (AR) Horizontal gaze dysfunction, sensorineural deafness, cerebrovascular anomalies, cardiac malformation, developmental delay, limb anomalies, autism. **Rx**: Team of specialists (pediatrics, surgery, cardiology, dental/speech/hearing) may be needed for specific symptoms.

Brachydactyly-ID (AD) Short digits. **Rx**: Plastic surgery/physical therapy in some.

Brain and spinal cord cancers (Increased risk in certain syndromes, e.g. neurofibromatosis, tuberous sclerosis, Li-Fraumeni syndrome, von Hippel-Lindau syndrome). **Rx**: Surgery, chemo/radiotherapy, targeted therapy.

Brain-lung-thyroid syndrome (AD) Movement disorder (chorea, athetosis, ataxia, myoclonus), muscle weakness, hypothyroidism, respiratory distress syndrome, interstitial lung disease, pulmonary fibrosis, lung cancer susceptibility. **Rx**: Tetrabenazine/levodopa for chorea; thyroid replacement therapy; lung problems treated as needed.

Branchiootorean/branchiootic syndrome (AD) Malformations of ear and kidney. **Rx**: Surgery, kidney transplant/dialysis in some.

Breast cancer (Most not inherited, may be AD with increased tendency to occur). **Rx**: Surgery/chemotherapy/radiotherapy in some cases.

Brooke-Spiegler syndrome (AD) Multiple skin tumors from sweat glands, hair follicles, salivary glands. **Rx**: Excision, electrosurgery.

Burkitt lymphoma (AR, AD) Swelling, distortion of facial bones, enlarged lymph nodes from B lymphocyte tumor. **Rx**: Chemotherapy, rituximab antibody.

Campomelic dysplasia (AD, most are new mutations) Skeletal abnormalities, ambiguous genitalia, facial features, breathing difficulties. **Rx**: Respiratory assistance, orthopedic surgery, sex reassignment of male with female genitalia, genetic counseling.

Canavan disease (AR) Defect in maintenance of the myelin sheath, with defective neural transmission. Problems with motor/speech development, large head, irritability, feeding difficulty. **Rx**: Supportive, symptomatic, physical therapy/special education.

Carcinoid tumor (AD) Diarrhea, skin flushing, wheezing, fast heartbeat, high blood pressure in some, gastric hyperacidity. **Rx**: Surgery, chemotherapy, radiation.

Carney-Stratakis syndrome (AD) Neuroendocrine, gastrointestinal tumors. **Rx**: Surgery, embolization (to cut off tumor blood supply), radiotherapy, chemotherapy.

Carnitine uptake defect (AR) Encephalopathy, cardiomyopathy, muscle weakness. **Rx**: L-carnitine supplements.

Cerebral adrenoleukodystrophy (XR) Adrenocoretical insufficiency (weakness, weight loss, vomiting, coma); learning/behavior problems; vision/swallowing problems; poor coordination; dementia in some. **Rx**: Stem cell transplant, corticosteroid replacement, physical therapy, gene therapy.

Cerebro-oculo-facial skeletal syndrome (AR) Craniofacial and skeletal abnormalities, cognitive disability. **Rx**: Supportive, tube feeding, genetic counseling.

Charcot-Marie-Tooth disease (AD) Peripheral nerve damage; loss of sensory, motor function in legs, feet, hands. **Rx**: Physical/occupational therapy, orthopedic surgery, analgesics.

CHARGE syndrome (AD) **C**oloboma, **H**eart defects, **A**tresia choanae (narrowed nasal breathing passage), **R**etarded growth, **G**enital/urinary and **E**ar abnormalities. **Rx**: Multiple specialists (medical/surgical, physical/occupational therapy, special education for multiple problems).

Childhood acute lymphoblastic leukemia (AR, but generally non-inherited somatic mutation) Fever, infections, bleeding, joint/bone pain, paleness easy bruising, shortness of breath, excess white blood cells in bone marrow. **Rx**: Chemotherapy, immunotherapy, targeted therapy, stem cell transplant.

Cholesterol desmolase deficiency (AR) Adrenal insufficiency, underdeveloped genitalia. **Rx**: Hormone replacement.

CHOPS syndrome (AD, de novo germinal mutation with no family history) **C**ognitive impairment, **C**oarse facial features, **H**eart defects, **O**besity, **P**ulmonary disease, **S**hort stature, **S**keletal abnormalities. **Rx**: Symptomatic, tracheostomy if severe.

Choroideremia (XR) Progressive vision loss, mainly in males; night blindness; tunnel vision. **Rx**: Gene therapy.

Chorioretinal degeneration (AD) Visual loss. **Rx**: Photocoagulation.

Chronic fatigue syndrome (Mulifactorial) **Rx**: Counseling, exercise.

Chronic granulomatous disease (XR or AR) Chronic inflammation, bacterial and fungal infections. **Rx**: Antibiotic/antifungal medication, corticosteroids, interferon gamma-1b.

Citrullinemia (AR) Ammonia accumulation, lethargy, feeding difficulty, liver problems. **Rx**: Medication to remove blood ammonia, dialysis, low-protein diet. Liver transplantation in some.

Claes-Jensen XLMR (XR) Intellectual disability, facial dysmorphism, undescended testes. **Rx**: Family counseling.

Cockayne syndrome (AR) Microcephaly, failure to thrive, short stature, delayed development, photosensitivity, hearing/vision loss, bone abnormalities. **Rx**: Gastrostomy tube if failure to thrive, hearing aids, cataract evaluation, antiseizure/antispasticity medication in some.

Coffin-Lowry syndrome (XD) Intellectual disability, delayed development, microcephaly, collapse when startled. **Rx**: Symptomatic, physical/speech/special education therapy.

Coffin-Siris syndrome (AD) Developmental disability, abnormal facial features and fifth fingers/toes, nail hypoplasia. **Rx**: Symptomatic, physical/occupational/speech/special education therapy.

Colorectal cancer (AR, AD or sporadic). **Rx**: Surgery; radiation therapy/chemotherapy in some.

Cone-rod dystrophy (AD, AR, XR) Vision loss. **Rx**: Light avoidance, low-vision aids.

Congenital adrenal hyperplasia (AR) Salt wasting, decreased fertility, weight loss, genital anomalies. **Rx**: Steroid replacement.

Congenital central hypoventilation syndrome (AD) Congenital stoppage of breathing, especially in sleep; problem regulating heart rate/blood pressure; some may have obstructed large colon (Hirschsprung disease); learning difficulties and other neurological problems, susceptibility to developing nervous system tumors. **Rx**: Positive pressure ventilation through tracheostomy.

Congenital deafness (AR, AD) **Rx**: Cochlear implants.

Congenital hypothyroidism (AR) **Rx**: Thyroid hormone replacement.

Coproporphyrinogen oxidase deficiency (variegate porphyria) (AD) Intermittent body pain in abdomen/back, nausea, vomiting, hypertension, tachycardia, orthostatic hypotension, seizures/peripheral neuropathy/muscle weakness in some. **Rx**: Symptomatic, panhematin (limits synthesis of porphyrin) to suppress acute attack of porphyria; liver transplant in some.

Cornelia de Lange syndrome (AD, XD) Slow growth; short stature; intellectual disability; abnormalities of facial features, bones in arms, hands, and fingers; excess body hair; visual impairment; other anomalies. **Rx**: Symptomatic, surgery for cleft palate/cardiac defects/diaphragmatic hernias; plastic surgery for excess hair.

Cori disease (Deficiency of debranching enzyme; Type III GSD) (AR) Hypoglycemia, hyperlipidemia, hepatomegaly, cirrhosis, slow growth, short stature, adenomas in some, muscle weakness. **Rx**: High-protein diet with cornstarch supplementation (cornstarch is digested slowly and can maintain normal blood sugar levels), liver transplant in some with liver cancer.

Cowden syndrome (AD) Multiple noncancerous tumors (hamartomas), increased cancer risk. **Rx**: Monitor for tumors.

Cranioectodermal dysplasia (AR) Bone abnormalities, facial features, missing teeth, kidney stones, cardiac anomalies. **Rx**: Group support, corrective surgery, speech/physical educational therapy, low-vision aids.

Craniofacial-deafness-hand syndrome (AD) Facial features, hearing loss, ulnar deviation of fingers with contractures. **Rx**: Surgical release of contractures.

Creutzfeldt-Jakob disease (AD) Dementia, ataxia. **Rx**: Supportive care, analgesics.

Cri du chat syndrome (Mostly not inherited; random deletion in chromosome 5 in reproductive cells, or inherited unbalanced translocation) High-pitched cat-like cry, intellectual disability, delayed development, microcephaly, hypotonia, facial features, some with heart defect. **Rx**: Physical/speech therapy.

Crigler-Najjar syndrome (AR) Cannot clear bilirubin; jaundice, bilirubin encephalopathy, muscle spasms, opisthotonus (spastic posture), hearing loss. **Rx**: Phototherapy, treat infections, plasmapheresis to lower bilirubin levels, liver transplant in some.

Critical Congenital Heart Disease (Mostly sporadic) **Rx**: Surgery or medication, depending on findings.

Crohn disease (Multifactorial) Intestinal inflammation. **Rx**: Anti-inflammatory medications, antibiotics, diet; surgery in some.

Cystathioninuria (deficiency of cystathioninase) (AR) Accumulation of plasma cystathionine; intellectual disability in some cases. **Rx**: Vitamin B6 in some.

Cystic fibrosis (AR) Respiratory, digestive problems from thick, sticky mucus. **Rx**: Antibiotics, loosening and removing mucus from lungs, treating intestinal blockage, good nutrition.

Cystinuria (AR) Urinary cystine stones. **Rx**: Remove stones, urinary alkalinization, thiol drugs to reduce cystine stones.

Dentatorubral-pallidolusian atrophy (AD) Involuntary movements, emotional and thinking problems. **Rx**: Anti-epileptic drugs.

Diabetes mellitus and deafness (Mitochondrial) Hyperglycemia, increased thirst/hunger/urination, mood changes, blurred vision, fatigue/weakness, sensorineural hearing loss. **Rx**: Dietary changes, hypoglycemic agents, insulin if needed.

Diabetes mellitus Type I (Multifactorial) Hyperglycemia, increased thirst/hunger/urination, mood changes, blurred vision, fatigue/weakness. **Rx**: Insulin, diet, exercise, weight control.

Diabetes mellitus Type II (Multifactorial) Hyperglycemia, increased thirst/hunger/urination, mood changes, blurred vision, fatigue/weakness. **Rx**: Hypoglycemic drugs, insulin, diet, exercise, weight control.

Diamond-Blackfan anemia (AD) Bone marrow dysfunction leads to anemia, with weakness, fatigue, pallor; short stature; facial features; risk for acute myeloid leukemia, osteosarcoma, and other cancers; cataracts, glaucoma, urogenital problems in some. **Rx**: Corticosteroids, blood transfusion, chelation therapy, bone marrow transplant.

DiGeorge syndrome (AD) Heart abnormalities, facial features, immune system deficiency, infections, cleft palate, breathing problems, kidney abnormalities, seizures, thrombocytopenia, GI problems, hearing loss. **Rx**: Surgery for heart defect, cleft palate, physiotherapy; mental health care.

Diphosphoglyceromutase deficiency (AR) Anemia. **Rx**: Blood cell transfusion in some cases.

DIRAS3 breast and ovarian cancer (Maternal imprinting) DIRAS3 gene, a tumor suppressor, is not expressed in breast and ovarian cancer. **Rx**: Chemotherapy, surgery.

Down syndrome (Trisomy 21) Intellectual disability; characteristic face; heart, gastrointestinal defects; happy demeanor. **Rx**: Physical/occupational therapy, special education, assistive aids (hearing aids, pencil grips, walking aids, seat cushions, touchscreen tablets).

Down syndrome with acute megakaryoblastic leukemia. **Rx**: Anti-tumor agents.

Duane-radial ray syndrome (AD) Eye movement problem, bone abnormalities in hands, other malformations, short stature. **Rx**: Surgery for ocular strabismus, malformations of hands/forearms, heart defects; hearing aids; growth hormone therapy in some.

Dubin-Johnson syndrome (AR) Jaundice, weakness. **Rx**: No specific treatment needed as symptoms are mild.

Dyskeratosis congenita (XR, AD, AR) Abnormal finger/toenails, skin pigmentary changes, white patches in mouth. **Rx**: Stem cell transplant.

Ehlers-Danlos syndrome (AD or de novo mutation) Hypermobile joints, elastic skin; internal bleeding in some cases; spinal curvature in some. **Rx**: Physical therapy, surgery, analgesics if needed.

Ellis-van Creveld syndrome (AR) Short limbs, extra fingers/toes, abnormal fingernails/teeth, congenital heart defects. **Rx**: Team approach for specific problems. Surgery, cardiology, dentistry, pulmonology, orthopedist as needed. Genetic counseling.

Enolase deficiency (AD) Myalgia, weakness, exercise intolerance. **Rx**: No specific treatment.

Epidermodysplasia verruciformis (XR, AR) Skin lesions, skin cancer risk. **Rx**: Remove lesions surgically or chemically; limit sun exposure.

Epidermolysis bullosa (AR, AD) Skin blistering, sometimes in lining of mouth or stomach. **Rx**: Avoid skin damage/infections; skin grafting; surgical widening of esophagus in some.

Essential fructosuria (fructokinase deficiency) (AR) Fructose loss in urine. **Rx**: No treatment indicated.

Essential pentosuria (xylitol dehydrogenase deficiency) (AR) High urine levels of L-xylulose. **Rx**: No treatment indicated.

Ewing sarcoma (Somatic mutation) Bone tumor. **Rx**: Chemotherapy, stem cell transplantation, surgery, radiation.

Fabry disease (X-linked partial penetrance). Lysosome storage disease. Pain in hands and feet, red spots between navel and knees, vision/hearing loss, decreased sweating, stomach pain, kidney disease in some. **Rx**: Restore missing enzyme alpha-galactosidase A; oral migalastat hydrochoride, a chaperone that helps move alpha-galactosidase A to lysosomes; pain management, avoid strenuous exercise or temperature extremes; low-protein/low-sodium diet if kidney dysfunction.

Fallopian tube cancer (10-15% hereditary AD increased susceptibility). **Rx**: Surgery, adjuvant chemotherapy.

Familial adenomatous polyposis (AD). Numerous large intestine polyps with malignant potential. May begin with multiple benign polyps in the teenage years. **Rx**: Surgical removal of colon.

Familial breast and ovarian cancer (AD increased susceptibility 38-84%; most sporadic). **Rx**: Counseling regarding prophylactic bilateral mastectomy/salpingo-oophorectomy; chemoprevention with tamoxifen in some.

Familial combined hyperlipidemia (AD) High cholesterol/triglyceride levels. Early heart attacks. **Rx**: Lower saturated fat intake, eat more fiber, avoid high-cholesterol foods; cholesterol-lowering medication; fish oil supplements.

Familial cylindromatosis (AD) Multiple skin tumors. **Rx**: Surgery/cryotherapy/laser therapy.

Familial dysautonomia (AR) Autonomic nervous system dysfunction. Feeding difficulties, growth delay, awkward gait, lack of tearing, labile body temperature/blood pressure, pulmonary infections, visual loss from corneal opacities and optic neuropathy, loss of bladder control; scoliosis. **Rx**: Feeding tube in some, artificial tears, medication for hypertension, vomiting, and bladder control; surgery for scoliosis if needed; counseling.

Familial dysbetalipoproteinemia (AR, AD) Increased total cholesterol/triglyceride levels. Fatty acid deposits (xanthomas) on eyelids, palms of hands, soles of feet and knee/elbow tendons; angina/coronary artery disease at young age. **Rx**: Statins, fibrates to lower triglycerides.

Familial hypercholesterolemia (AD, AR). Risk for heart disease. Cholesterol deposition in tendons (xanthomas), eyelids (xanthelasmata), and cornea (arcus cornealis). **Rx**: Lipid-lowering drugs, low saturated fat diet, fiber in diet, exercise, weight control.

Familial hyperinsulinism (AR, AD) Abnormally high insulin levels. Hypoglycemia, lethargy, feeding difficulty. **Rx**: Dietary modification, glucose administration, medication to reduce insulin release, surgery to remove part of pancreas if needed.

Familial hypertriglyceridemia (AD) Xanthomata (yellow skin nodules from lipid deposition), hepatosplenomegaly, risk of pancreatitis. **Rx**: Weight loss, exercise, fibrates, statins in some.

Familial hypobetalipoproteinemia (Autosomal codominant – genes from both parents are expressed) Low cholesterol, fatty liver, cirrhosis, excess fat in feces, failure to thrive. Deficiency of fat-soluble vitamins (D, E, A, K), clotting problems in some. **Rx**: Vitamin supplementation in some.

Familial lipoprotein lipase deficiency (AR) Abdominal pain, pancreatitis, xanthomas, hepatosplenomegaly. **Rx**: Dietary fat restriction, avoid alcohol and drugs that increase triglyceride levels; gene therapy.

Familial Mediterranean fever (AR, AD) Intermittent fever; pain in abdomen, chest, and joints. **Rx**: Symptomatic - IV saline for hydration, analgesics.

Familial nonchromaffin paraganglioma (AD) Vascularized, usually benign tumors of head, neck, carotid body. **Rx**: Surgery, embolization to cut off tumor circulation, radiotherapy.

Fanconi anemia (AR, rarely XR) Aplastic anemia, low white cells, platelets, tiredness, infections, skin hypopigmentation/cafe au lait spots, physical abnormalities, poor growth, short stature. **Rx**: Stem cell transplant, androgen therapy, synthetic growth factors, gene therapy.

Fanconi syndrome (AR) Kidney proximal tubule reabsorption problems, with failure to thrive, excess thirst/urination, rickets, bone disease, weakness, low blood phosphate, potassium, loss of amino acids in urine, corneal abnormalities. **Rx**: Fluid/electrolyte replacement, vitamin D, high-calorie diet.

Fascioscapulohumeral muscular dystrophy (AD) Muscle weakness/wasting, particularly in face, shoulder blades and upper arms. **Rx**: Anti-inflammatory drugs for comfort, surgery to stabilize shoulder blades, orthopedic supports.

Ferrochelatase (heme synthetase) deficiency (AD, AR) Overproduction of porphyrins with photosensitivity, skin irritation, nail deformity, weakness, fatigue, dyspnea on exertion, orthopnea, palpitations, fever. **Rx**: Sunscreen, avoid sun exposure; beta carotene for photosensitivity; cholestyramine to remove protoporphyrin from enterohepatic circulation; hemolysis/splenomegaly/liver transplant for marked symptoms.

Floating-Harbor syndrome (AD) Short stature; delayed bone, speech development; facial features. **Rx**: Supportive – auditory/visual, orthopedic, dental, family counseling.

Formiminotransferase deficiency (AR) Intellectual disability, delayed development, elevated blood folate, high urine levels of formiminoglutamate, megaloblastic anemia in some, tingling/numbness of hands. **Rx**: Supportive.

Fragile X syndrome (XD) Learning, speech, cognitive impairment, facial features, large testes. **Rx**: Symptomatic, physical/occupational/speech therapy, special education, special computer apps.

Friedreich ataxia (AR) Ataxia, decreased strength and sensation in arms and legs, impaired speech, hearing, vision, hypertrophic cardiomyopathy, scoliosis. **Rx**: Physical/occupational therapy, surgery/braces for bone deformity, medication for heart disease.

Frontonasal dysplasia (AR) Abnormal head shape, absence of brain corpus callosum. **Rx**: Surgery for deformities, genetic counseling, social/educational services.

Fumarase deficiency (AR) Developmental delay, microcephaly, failure to thrive, seizures, facial features. **Rx**: Antiseizure medication, feeding tube if needed, physical/speech therapy.

Galactosemia, classic (AR) Failure to thrive, liver damage, abnormal bleeding, bacterial infections, cataracts, intellectual disability. **Rx**: Low-galactose diet (cataracts are then reversible).

Gastric cancer (AD). **Rx**: Surgery, chemotherapy, radiation.

Galactokinase deficiency (defect in the galactose 1-P-uridyl transferase) (AR) Cataracts. **Rx**: Decrease lactose/galactose in diet (cataracts are then often reversible).

Gardner syndrome (AD) Multiple colorectal polyps and various benign/malignant tumors. Risk of early age colorectal cancer. **Rx**: Regular screening, surgery, chemo/radiation therapy.

Gaucher disease (AR) Liver/spleen enlargement, anemia, easy bruising, bone abnormalities, brain damage, skin abnormalities, facial features. **Rx**: Enzyme replacement, orthopedic surgery, blood transfusions, pain medication if needed.

Genitopatellar syndrome (AD) Genital abnormalities, underdeveloped patellae, clubfoot, delayed development, thyroid anomalies, intellectual disability, microcephaly, absent corpus callosum. **Rx**: Educational/speech therapy, orthopedic surgery for some, physical therapy, thyroid hormone replacement.

Gilbert syndrome (deficiency in UPD-glucuronyl transferase) (AD) Mild liver condition, jaundice. **Rx**: No treatment needed.

Glucose-6-phosphate dehydrogenase deficiency (XR) Hemolytic anemia, jaundice, fatigue, shortness of breath, large spleen. **Rx**: Avoid hemolysis triggers (infections, fava beans, certain drugs).

Glucose-6-phosphatase deficiency (Type I GSD; Von Gierke's Disease) (AR) Growth retardation, enlarged liver from glycogen/fat accumulation in liver, hypoglycemia, tendency for infections, risk of liver tumors, inflammatory bowel disease, renal insufficiency. **Rx**: Diet to avoid hypoglycemia; liver/ kidney transplant in some.

Glucose/galactose malabsorption (AR) Dehydration, acid blood, weight loss, kidney stones. **Rx**: Eliminate milk products from diet.

Glucose phosphate isomerase deficiency (AR) Hemolytic anemia, jaundice, fatigue, shortness of breath, tachycardia, enlarged spleen, gallstones. Intellectual disability and ataxia in some. **Rx**: Supportive, transfusions/splenectomy in some.

Glutaric acidemia Type I (AR) Can›t break down lysine, hydroxylysine, and tryptophan; macrocephaly; movement difficulty; bleeding in brain or eyes; carnitine deficiency. **Rx**: Low-lysine diet, carnitine supplementation to remove glutaric acid.

Gorlin syndrome (AD) Risk for basal cell carcinomas, other tumors; facial features, skeletal abnormalities. **Rx**: Surgery.

Gout (Multifactorial) Joint inflammation, high uric acid in blood, kidney stones in some, hypertension. **Rx**: Anti-inflammatory drugs, drugs that reduce uric acid production.

Gray baby syndrome (underdeveloped microsomal UDP-glucuronyl transferase) (AR) Cyanosis, hypotension, blue lips and skin, difficulty breathing, irregular heartbeat, exacerbated by chloramphenicol. **Rx**: Stop chloramphenicol, exchange blood transfusion/hemodialysis/oxygen therapy in some.

Hand-foot-genital syndrome (AD) Short thumb/5th finger/1st toe/feet, fused wrist/ankle bones, genital/urinary tract abnormalities. **Rx**: Bladder/ureter surgical correction.

Hartnup disease (AR) Error in amino acid metabolism with high urine tryptophan and other amino acids; can't produce enough niacin; central nervous system abnormalities; rash; short stature; headache; fainting. **Rx**: Niacinamide/niacin, avoid sun exposure.

Hb S/beta-thalassemia (AR) Combines features of sickle cell disease (cell sickling) with thalassemia (reduced mature red cells); anemia; infections; pain. **Rx**: Drink lots of water; avoid extreme climate change; blood transfusion; hydroxyurea, a myelosuppressive agent, to raise the level of fetal hemoglobin (HbF).

Hemochromatosis (AR) Body absorbs too much iron, which is stored in and damages the liver and pancreas. Fatigue, joint/abdominal pain, cirrhosis, liver cancer, diabetes, heart problems, skin discoloration, decreased sex drive, memory problems, hair loss. **Rx**: Reduce the amount of iron in the body through chelation therapy, phlebotomy, dietary modification; treatment of complications.

Hemophilia A & B (XR) Continuous bleeding from minor trauma. Type A = factor VII deficiency; Type B = factor IX deficiency. **Rx**: Provide clotting factor.

Hepatoblastoma (Somatic mutation) Liver cancer; associated with Beckwith-Wiedemann syndrome (AD), familial adenomatous polyposis, and Aicardi syndrome. **Rx**: Chemotherapy, surgical resection, radiation therapy.

Hereditary elliptocytosis (AD) Also called **hereditary ovalocytosis**. Fatigue, shortness of breath, gallstones, jaundice, enlarged spleen. **Rx**: Transfusion/splenectomy if severe.

Hereditary fructose intolerance (deficiency in fructose 1-phosphate aldolase) (AR) Failure to thrive, fructose intolerance (nausea, abdominal pain, diarrhea, vomiting, hypoglycemia). **Rx**: Eliminate fructose and sucrose from diet.

Hereditary nonpolyposis colorectal cancer (AD) Also called **Lynch syndrome**. Risk of colon/rectal and other gastrointestinal cancers and other cancers. **Rx**: Removal of colon.

Hereditary ovalocytosis (AD) Also called **hereditary elliptocytosis**. Fatigue, shortness of breath, gallstones, jaundice, enlarged spleen. **Rx**: Transfusion/splenectomy if severe.

Hereditary spherocytosis (AD, AR) Anemia, jaundice, enlarged spleen, gallstones. **Rx**: Blood transfusion, folic acid, splenectomy, cholecystectomy in some.

Hereditary stomatocytosis (AD) Abnormal red cell permeability. Hemolytic anemia, splenomegaly, gallstones. **Rx**: Folate administration, red cell transfusion, chelating therapy for iron overload.

Hexokinase deficiency (AR) Hemolytic anemia, jaundice, splenomegaly, lethargy, fatigue. **Rx**: Blood transfusions in some.

Hirschsprung disease (AD) Absent nerves from part of intestine; enlarged, often inflamed colon. **Rx**: Removal of diseased part of colon.

Histidinemia (lack of histidase) (AR) Elevated blood histamine. **Rx**: No treatment indicated.

Hodgkin disease, Y-linked pseudoautosomal (YD) Lymph node cancer. **Rx**: Radiotherapy, chemotherapy.

Holt-Oram syndrome (AD) Abnormal development of upper limb bones and heart. May have absent arm, wrist, and thumb bones, as well as cardiac atrial or ventricular septal defects and cardiac arrhythmias. **Rx**: Surgical correction of heart/limb abnormalities; medication/pacemaker if needed for cardiac arrhythmias; limb prosthetics; physical/occupational therapy.

Homocystinuria (AR) Myopia, dislocated lens, abnormal blood clotting, osteoporosis, developmental delay, anemia. **Rx**: High doses of vitamin B6 help in about 50%; diet low in methionine; betaine removes homocysteine from the blood.

Hunter syndrome (deficiency in lysosomal iduronate-2-sulfatase) (XR) Facial features, hoarse voice, narrow airway, sleep apnea, macrocephaly, hydrocephalus, hepatosplenomegaly, umbilical hernia, thick skin, hearing loss, ear infections, retina dysfunction, carpal tunnel syndrome, spinal stenosis, heart valve problems, short stature, skeletal abnormalities, intellectual dysfunction in some cases. **Rx**: Management of symptoms and complications; enzyme replacement; bone marrow/umbilical cord blood stem cell transplant.

Huntington disease (AD) Progressive brain disorder with uncontrolled movements, cognitive difficulty. Symptoms usually appear between ages 30-50 and progress over 10-25 years. **Rx**: Symptomatic; antipsychotic drugs, antidepressants, tranquilizers, exercise; tetrabenzine decreases dopamine/serotonin and reduces abnormal movements. Gene therapy trials underway.

Hurler disease (AR) A lysosomal storage disease. Facial features, spleno/hepatomegaly, developmental delay, skeletal deformities, cardiomyopathy, upper airway obstruction, hearing loss, corneal opacity. **Rx**: Hematopoetic stem cell transplant, enzyme replacement therapy (laronidase), supportive care, physical/occupational therapy, speech therapy, respiratory support, hearing aids, pain medication, orthopedic correction.

Hutchinson-Gillford progeria syndrome (AD) Premature aging. **Rx**: Low-dose aspirin, physical/occupational therapy, farnesyltransferase inhibitors may help with weight gain and increased blood vessel flexibility.

Hydroxymethylglutaryl lyase deficiency (AR) Cannot process leucine. Dehydration, lethargy, hypotonia, hypoglycemia, metabolic acidosis, breathing problems, seizures; carnitine deficiency. **Rx**: Low-leucine/high-carbohydrate diet, dietary supplementation with L-carnitine, avoid alcohol, reduce blood ammonia with sodium phenylacetate and sodium benzoate.

Hypercholesterolemia (AD) Heart disease, xanthomas, arcus cornealis. **Rx**: Diet, exercise, statins, inhibitors of cholesterol absorption.

Hyperkalemic periodic paralysis (AD) Episodes of extreme weakness starting at early age. **Rx**: High-carbohydrate foods; diuretics for acute attack.

Hypermethioninuria (decrease in methionine adenosyl transferase) (AR) Increased blood methionine. Intellectual disability, other neurological problems, delayed motor skills, weakness, facial features, smell of boiled cabbage. **Rx**: Methionine-restricted diet; S-adenosylmethionine supplementation.

Hyperparathyroidism, familial (AD) Kidney stones, hypertension, weakness, fatigue, osteoporosis. **Rx**: Parathyroidectomy; calcimimetic drugs to trick parathyroid into producing less parathyroid hormone; hormone replacement for osteoporosis, bisphosphonates to prevent calcium loss from bones.

Hypervalinemia (defect in valine transaminase) (AR) High levels of valine in blood/urine, failure to thrive, vomiting, poor appetite, drowsiness, developmental delay. **Rx**: Low-valine diet.

ICF syndrome (AR) (**I**mmunodeficiency-**C**entromeric instability-**F**acial anomalies) Growth/psychomotor retardation, low immunoglobulins, infection, facial anomalies. **Rx**: Immunoglobulin infusion, stem cell transplant.

Immunodeficiency with hyper-IgM (XR) Susceptibility to infections, cancer. **Rx**: Immunoglobulin replacement, antibiotics, hematopoetic stem cell transplant in some.

Isovaleric acidemia (AR) Poor feeding, seizures, lethargy, sweaty foot odor; carnitine deficiency. **Rx**: Protein-restricted diet; oral glycine and L-carnitine to reduce excess isovaleric acid.

Ivic syndrome (AD) Upper limb abnormalities, ocular motor disturbances, hearing loss. **Rx**: Manage signs and symptoms.

Joubert syndrome (AR, XR) Brain abnormalities, ataxia, hyperpnea/apnea, abnormal eye movements, delayed development, intellectual disability, facial features, coloboma, kidney disease. **Rx**: Symptomatic, physical/occupational/speech therapy.

Kabuki disease (AD, XD) Facial features, developmental delay, seizures, microcephaly, hypotonia, nystagmus, scoliosis, short pinky, other skeletal abnormalities, heart problems, ear infections, hearing loss, early puberty. **Rx**: Physical/occupational/speech therapy, special education, sensory stimulation.

Kartagener syndrome (AR) Ciliary disorder with situs inversus, chronic sinusitis, bronchiectasis, chest infections, infertility. **Rx**: Antibiotics, immunizations, in vitro fertilization.

Kidney cancer (AD) **Rx**: Surgery, targeted therapy, immunotherapy, radio/chemotherapy.

Kleefstra syndrome (AD) Developmental delay, autism, limited speech, hypotonia, microcephaly, brachycephaly, seizures, facial features, brain abnormalities, heart defects, genitourinary problems. **Rx**: Special education, speech/physical/occupational/vocational/sensory stimulation therapy.

Klinefelter syndrome (Extra X chromosome—XXY) Tall, infertility, decreased testosterone, may have delayed puberty, gynecomastia, decreased bone density, decreased body hair, other physical changes, impaired social skills. **Rx**: Testosterone replacement, breast tissue removal, speech/physical therapy, fertility treatment (intracytoplasmic sperm injection), psychological counseling.

Krabbe disease (Globoid leukodystrophy) (AR) Failure to thrive, rigid muscle tone, weakness, ataxia, hearing/vision loss, feeding difficulty, sensitivity to loud sounds, fever. **Rx**: Supportive care, stem cell transplants successful in some cases.

Lactase deficiency (AR) Can't digest lactose; diarrhea, bloating, gas after eating dairy. **Rx**: Reduce dietary lactose; enzyme replacement with lactase tablets helps some.

Langer mesomelic dysplasia (Pseudoautosomal recessive X and Y chromosome) Bone growth problem, shortened long bones of arms and legs, short stature. **Rx**: Orthopedic surgery in some.

LCAT (Lecithin-Cholesterol AcylTransferase) Deficiency (AR) Cholesterol accumulation. Corneal cloudiness, kidney disease, hemolytic anemia, hepato/splenomegaly in some, atherosclerosis. **Rx**: Corneal transplant, kidney dialysis/ transplant in some cases.

Leber congenital amaurosis (AR, AD) Retinal vision impairment since infancy, photophobia, nystagmus, hyperopia. **Rx**: Gene therapy, low-vision aids, mobility training.

Leber hereditary optic neuropathy (Mitochondrial inheritance) Progressive cloudy vision beginning in teens or twenties, sometimes movement disorders. **Rx**: Idebenone is thought to improve energy production in mitochondria; avoid alcohol/smoking; low-vision aids.

Leigh syndrome (AR, XR, mitochondrial in 20%) Severe mental/movement problems in infancy, usually death in 2-3 yrs. **Rx**: Thiamine or vitamin B1, oral sodium bicarbonate or sodium citrate for lactic acidosis.

Leri-Weill dyschondrosteosis (XD, pseudoautosomal YD) Bone growth disorder, short stature, muscle hypertrophy. **Rx**: Supportive, counseling, growth hormone.

Lesch-Nyhan syndrome (Deficiency of hypoxanthine P-ribosyl transferase) (XR) Kidney dysfunction, gouty arthritis, self-mutilating behaviour (lip/finger biting, head banging), involuntary movements and other neurologic problems. **Rx**: Uric acid-lowering drugs, shock wave lithotripsy for kidney stones; antispastic drugs; behavior modification treatment, genetic counseling.

Leukemia (AR, AD, XR, or somatic mutation). **Rx**: Chemo/radiation therapy, targeted therapy, immunotherapy, stem cell transplant. Splenectomy in some with enlarged spleen.

Li-Fraumeni syndrome (AD) Risk of developing multiple kinds of cancer. **Rx**: Surgery/chemotherapy.

Liddle syndrome (AD) Severe hypertension since childhood, hypokalemia. **Rx**: Low-sodium diet, potassium-sparing diuretics. (The usual anti-hypertensive medications not effective)

Liebenberg syndrome (AD) Malformation of arms and bone of other extremities. **Rx**: Corrective surgery.

Ligase 4 syndrome (AR) Immunodeficiency, microcephaly, growth failure, learning problems, susceptible to lymphoid malignancy. **Rx**: Supportive, antibiotics, immunoglobulins, stem cell transplant.

Liver phosphorylase deficiency (Type VI GSD; Hers Disease) (AR) Generally mild, but with hepatomegaly and growth retardation, hyperlipidemia; hyperketosis, hypoglycemia when fasting. **Rx**: Avoid fasting; frequent small meals; uncooked cornstarch may help some.

Liver phosphorylase kinase deficiency (Type VIII GSD) (AR, XR) Hepatomegaly, growth retardation, hypercholesterolemia, hypertriglyceridemia, fasting hyperketosis. **Rx**: Frequent small meals of carbohydrates and cornstarch.

Long-chain hydroxyacyl-CoA dehydrogenase deficiency (AR) Can't break down certain fats. Feeding problems, lethargy, hypoglycemia, hypotonia, liver and retinal problems, muscle pain. **Rx**: Medium chain triglyceride oil, which body can break down.

Lou Gehrig disease (Amyotrophic lateral sclerosis) (5-10% inherited, AD, AS, XD) Muscle atrophy, weakness, hyperreflexia. **Rx**: Riluzole, to decrease damage to motor neurons by minimizing glutamate release; physical therapy.

Lujan syndrome (XR) Intellectual/behavioral difficulties, macrocephaly, facial features, hyperextensibility, other anomalies. **Rx**: Special education, treatment of scoliosis.

Lupus erythematosus (Multifactorial) Arthritis, fever, fatigue, skin rash (butterfly rash on nose and cheeks), kidney damage, pleurisy, pericarditis, infection. **Rx:** Nonsteroidal anti-inflammatory agents, corticosteroids, antimalarial drugs, immunosuppressives, other drugs.

Lynch syndrome (AD) Risk of colon/rectal and other gastrointestinal cancers and other cancers (also called **hereditary nonpolyposis colorectal cancer**). **Rx**: Removal of colon.

Male infertility (Multiple causes including AR, XR) **Rx**: Surgery for anatomical abnormalities, treat infections, hormone treatment, assisted reproductive technology (in vitro fertilization, intracytoplasmic sperm injection, cryopreservation of gametes or embryos, fertility medication).

Maple syrup urine disease (AR) Sweaty urine odor, poor feeding, lethargy, abnormal movements, delayed development, seizures. **Rx**: Dietary restriction of branched-chain amino acids, vitamin B1 administration, liver transplant in some.

Marfan syndrome (AD, 25% spontaneous mutation) Connective tissue abnormality affecting strength/flexibility of bones, ligaments, muscles, blood vessels (aortic aneurysm/dissection) and heart valves; tall, slender, large arm span; dislocated lens. **Rx**: Blood pressure lowering medication, surgery for aortic repair, vision correction, scoliosis treatment.

Maturity onset diabetes of the young Type 5 (AD) Pancreas does not make enough insulin. More likely to occur in adolescents and young adults. Hyperglycemia, infections, blurry vision, frequent urination/thirst, weight loss, kidney damage. **Rx**: Treat hyperglycemia and kidney problems.

Mayer-Rokitansky-Kuster-Hauser syndrome (AR in some) Vagina/uterus underdevelopment; sometimes kidney, skeletal, hearing, heart abnormalities. **Rx**: Surgical reconstruction.

McCune-Albright syndrome (Not inherited; somatic mosaicism) Bone scars, fractures, uneven bone growth, early puberty and other endocrine disorders, cafe au lait spots. **Rx**: Drugs to inhibit bone resorption, orthopedic correction.

McKusick-Kaufman syndrome (AR) Extra fingers, heart defects, genital abnormalities. **Rx**: Surgical correction.

Meckel-Gruber syndrome (AR) Enlarged, cystic kidneys; protrusion of brain through skull; extra fingers and toes; liver scars. **Rx**: Symptomatic, supportive, genetic counseling.

Medium-chain acyl-CoA dehydrogenase deficiency (AR) lethargy in infancy/early childhood, hypoglycemia, seizures, breathing problems, liver and brain damage. **Rx**: Frequent feeding, especially carbohydrates; avoid alcohol.

Medullary thyroid cancer, familial (AD) **Rx**: Thyroidectomy, immunotherapy.

Melanoma, hereditary (AD) Increased risk of pancreatic cancer in some. **Rx**: Sunscreen, avoid sun, lesion removal; chemo/radiotherapy, targeted therapy, immunotherapy.

MELAS (**M**itochondrial **E**ncephalomyopathy, **L**actic **A**cidosis and **S**troke-like episodes) **syndrome** (Mitochondrial) Usually starts in childhood. Muscle weakness/pain, headaches, decreased appetite, seizures, temporary hemiparesis, lactic acidosis, nitric acid deficiency, ataxia, hearing loss, other disability in some. **Rx**: Anti-convulsant drugs, cochlear implants in some, arginine and citrulline supplementation for nitric oxide deficiency.

Metachromatic leukodystrophy (AR) Myelin (white matter) destruction in nervous system. Intellectual/motor/sensory deterioration, behavioral problems. **Rx**: Medication for behavioral problems, seizures, pain, infection, GI problems; physical/occupational/speech therapy.

Methylmalonic aciduria (cblA and cblB forms) (AR) Dehydration in infant, hypotonia, developmental delay, lethargy, enlarged liver, failure to thrive, cognitive problems, kidney disease, pancreatitis; carnitine deficiency. **Rx**: Low-protein, high-calorie diet, oral antibiotics to reduce propionic acid from gut flora, vitamin B12, carnitine, organ transplantation in some cases.

Microduplication 17p13.3 syndrome. (Denovo, maternal inheritance) Psychomotor delay, dysmorphic features. **Rx**: Growth hormone for short stature; speech/occupational/physiotherapy.

Microduplication 22q11.2 syndrome (AD) Developmental delay, intellectual disability, short stature, muscle weakness, immune system defects, hearing loss, seizures, but many people have no physical/intellectual defect. **Rx**: Diverse specialists for specific problems.

Miller-Dieker syndrome (Most not inherited, may be AD) Decreased brain folds, intellectual disability, developmental delay, spasticity, seizures, hypotonia, facial features, kidney/gastrointestinal anomalies. Most do not survive childhood. **Rx**: Supportive, specialists for individual organ problems.

Mitochondrial DNA depletion syndrome (AR) Impaired energy production, hypotonia, muscle weakness, droopy eyelids, muscle weakness, breathing problems, sometimes epilepsy, liver failure. **Rx**: Physical therapy, drugs for epilepsy if present, liver transplant in some.

Mitochondrial myopathy (AR) Brain dysfunction, weakness, cardiomyopathy, liver dysfunction, difficulty swallowing; carnitine deficiency. **Rx**: Coenzyme Q10, B complex vitamins, riboflavin, alpha lipoic acid, L-carnitine, creatine, L-arginine, exercise.

Mowat-Wilson syndrome (AD) Intellectual disability, facial features, Hirschsprung megacolon, heart/kidney anomalies, happy demeanor. **Rx**: Physical/occupational/speech therapy. Surgery to remove diseased colon. Correct heart defects/urinary anomalies.

Mucolipidosis IV disease (AR) Delayed mental/motor development, cloudy corneas, retinal damage, anemia. **Rx**: Symptomatic, counseling. Physical/occupational/speech therapy; iron supplements if anemia; corneal transplantation.

Mucolipidosis VII disease (AR) A lysosomal storage disease. Growth retardation, skeletal abnormalities, facial features, some intellectual disability, hernias, osteoarthritis. **Rx**: Enzyme replacement, surgical correction of hernias and bone deformities.

Mucopolysaccharidosis VII (AR) Enlarged head/tongue, hydrocephalus, facial features, abdominal hernia, enlarged liver/spleen, cloudy cornea, hearing loss, intellectual disability in some. **Rx**: Enzyme replacement; surgical correction of bone deformities/inguinal/umbilical hernias; hematopoietic stem cell therapy.

Muir-Torre syndrome (AD) Sebaceous tumors, colorectal cancer, other tumors. **Rx**: Surgery; oral retinoids/ interferon injections to prevent skin lesions.

Multiple sulfatase deficiency (AR) Seizures, developmental delay, dry scaly skin, excess hair growth, skeletal abnormalities, facial features, movement problems. **Rx**: Supportive care for specific symptoms.

Muscle phosphorylase deficiency (Type V GSD; McArdle Disease) (AR) Exercise intolerance, muscle pain/ weakness/cramps. **Rx**: Moderate exercise, diet rich in carbohydrates, creatine supplements, vitamin B6.

Multiple endocrine neoplasia Types 2 and 3 (AD, new mutations) Multiple endocrine gland and other tumors, cancerous or malignant. **Rx**: Varies, depending on tumor.

Multiple familial trichoepithelioma (AD) Multiple skin tumors of hair follicles and sweat glands, usually noncancerous. **Rx**: Surgery, cryo/laser surgery.

Multiple sclerosis (May be family inheritance in some, but nature unknown) Considered an autoimmune disorder, scattered sensory/motor damage within nervous system, including vision problems, muscle control, weakness, bladder control. **Rx**: Corticosteroids, plasmapheresis, drugs to decrease progression by interacting with the immune system (monoclonal antibody, beta interferons, glatiramer acetate).

Multiple system atrophy (Sporadic, no clear inheritance) Progressive difficulty with movement and balance; autonomic dysfunction with blood pressure drop, urinary problems, erectile dysfunction. **Rx**: Medication to raise blood pressure; drugs for impotence, bladder control (botulinum toxin), posture; physical therapy.

Multiple-CoA carboxylase deficiency (AR) Feeding difficulty, lethargy, seizures, breathing problem, skin rash, developmental delay. **Rx**: Oral biotin.

Muscular dystrophy (XR) Multiple types. Muscle weakness and atrophy, particularly proximal skeletal muscles, maybe cardiomyopathy. **Rx**: Physical/respiratory/speech/occupational therapy. Corrective therapy. Corticosteroids, heart medication if needed; gene therapy; eteplirsen (skips exon 51), golodirsen (exon skipping), RNA interference.

Myoclonic epilepsy with ragged red fibers (Mitochondrial) Muscle weakness, twitching, spasticity, seizures, ataxia, decreased sensation in extremities, dementia, cardiomyopathy, carnitine deficiency, other abnormalities. **Rx**: Symptomatic treatment with Coenzyme Q10, B complex vitamins, L-carnitine; antiseizure medication, physical therapy.

Myoneurogenic gastrointestinal encephalopathy (AR) Gastrointestinal dysmotility, peripheral neuropathy, ptosis, ophthalmoplegia, hearing loss, deterioration of white matter in brain. **Rx**: Medication for nausea and vomiting, pain. Celiac plexus neurolysis for abdominal pain.

Myositis ossificans (AD) Bone replaces muscles/tendons/ligaments, loss of mobility, speaking and eating, malformed big toe. **Rx**: Anti-inflammatory medication, icing the injury, rest.

Myotonia congenita (AD, AS) Sustained muscle tensing without relaxation, usually legs. **Rx**: Exercise, muscle relaxants.

Myotonic dystrophy (AD) Muscle wasting, weakness, prolonged muscle contractions, can't release grip, cataracts, may have intellectual disability, clubfoot. **Rx**: Anti-myotonic/anti-inflammatory drugs, rehab, assistive devices.

Myotubular myopathy (XR) Muscle weakness since birth; feeding/breathing difficulty, impaired muscles of sitting, standing, walking, and, in some, eye movements and facial muscles; scoliosis. **Rx**: Physical/occupational therapy; respiratory/feeding tube support in some; gene therapy.

Nail-patella syndrome (AD) Nail abnormalities; dislocation of patella; at risk for glaucoma, kidney disease. **Rx**: Analgesics, physiotherapy, splinting; surgery in some.

Nephrogenic diabetes insipidus (XR, AR, AD) Body loses water through urine (polyuria); thirst. **Rx**: Hydrochlorothiazide diuretic, fluid replacement.

Nephronophthisis (AR) Inflammation and scarring of kidneys. **Rx**: Kidney transplant.

Neuroblastoma (AD) Tumor commonly in adrenal gland, usually in children, can metastasize. **Rx**: Chemotherapy, surgery.

Neuroendocrine tumors (AD) **Rx**: Surgery, chemotherapy, hormone therapy, targeted therapy.

Neurofibromatosis Types I and II (AD or spontaneous) Noncancerous tumors of nervous system Schwann cells, often in vestibular/acoustic nerves; hearing loss, ringing in ears, balance problems; extremity numbness/weakness. **Rx**: Surgery, radiation, chemotherapy.

Neuropathy, ataxia, retinitis pigmentosa, and ptosis (Mitochondrial) Muscle weakness, numbness, pain in arms and legs, ataxia, retinal degeneration. **Rx**: Symptomatic. Antioxidants improve oxidative phosphorylation.

Nicoladis-Baraitser syndrome (AD) Microcephaly, short stature, short fingers/toes, seizures, facial features, intellectual disability, happy demeanor. **Rx**: Symptomatic; physical/occupational/speech therapy.

Niemann-Pick disease (AR) Failure to thrive, hepatosplenomegaly, intellectual disability, lung disease, cherry-red spot in retina, short stature, early death. **Rx**: Supportive.

Nijmegen breakage syndrome (AR) Microcephaly, short stature, facial features, intellectual disability, pulmonary infections, increased cancer risk, immune deficiency. **Rx**: Vitamin E/folic acid supplements, prophylactic antibiotics, intravenous infusion with immunoglobulin in some for infections, bone marrow/stem cell transplants.

Nonsyndromic hearing loss (AR, AD, X-linked, mitochondrial) Hearing loss usually sensorineural (inner ear), sometimes conductive (middle ear). **Rx**: Hearing aids, cochlear implant.

Nonsyndromic holoprosencephaly (AD) Brains doesn't divide into two halves. Sometimes cyclopia, microcephaly, hydrocephalus, cleft palate, single front tooth, microphthalmia, developmental delay, intellectual disability, pituitary malfunction. **Rx**: Symptomatic, supportive.

Obesity (AD, AR, Multifactorial). **Rx**: Weight loss, diet, exercise.

Ocular melanoma and retinoblastoma (AD, most sporadic). **Rx**: Surgery, radiation, chemotherapy.

Oculocutaneous albinism (AR) Light-colored skin/iris. **Rx**: Sunglasses.

Oligopolyposis, multiple (AR) Colorectal adenomas. **Rx**: Surgery, chemotherapy, immune system stimulation.

Opitz-Kaveggia syndrome (XR) Intellectual disability, friendly persona, hypotonia, broad thumbs and first toes, imperforate anus, facial features, seizures, scoliosis, sensorineural hearing loss, heart abnormality. **Rx**: Address individual signs.

Ornithine transcarbamylase deficiency (XR) Ammonia accumulation, lethargy, poor eating/breathing, floppy, seizures, intellectual disability, developmental delay. **Rx**: Decrease protein intake, increase dietary carbohydrates and lipids, hemodialysis in comatose patients with high blood ammonia.

Orofaciodigital syndrome (XD, AR) Abnormal development of oral cavity (split tongue, benign tumors of tongue, extra or missing teeth, cleft palate), brain abnormalities, intellectual disability, fused fingers, extra digits, polycystic kidney disease in some. **Rx**: Reconstructive surgery, speech therapy, special education, treatment for renal disease and seizures.

Orotic aciduria (AR) Excess orotic acid in urine, failure to thrive, developmental delay, megaloblastic anemia. **Rx**: Uridine monophosphate or uridine triacetate to bypass the missing enzyme.

Osteogenesis imperfecta (AD, AR, XR) Fragile bones, multiple fractures. **Rx**: Prevention and care of fractures.

Ovarian cancer (AD but most are noninherited somatic mutations). **Rx**: Surgery, chemotherapy.

Ovarian insufficiency (~10% familial, multifactorial) Infertility. **Rx**: Estrogen therapy.

Pancreatic cancer (~10% hereditary; multifactorial). **Rx**: Surgery, radiation, chemotherapy, targeted therapy, immunotherapy.

Paraganglioma, hereditary (AD) Mostly benign tumors of the autonomic nervous system ganglia. Pheochromocytomas secrete epinephrine, with episodes of hypertension, and cardiac palpitations. **Rx**: Stabilize blood pressure, surgery, radiation, thermal ablation.

Parathyroid gland cancer (Not inherited, but AD may cause a predisposition) Hypercalcemia, kidney stones, bone fractures, difficulty swallowing. **Rx**: Surgery, radiation.

Parkinson disease (AD, AR in some; <10% believed due to genetic causes) Slowness, stiffness, tremor, dementia in some. **Rx**: Carbidopa-levodopa, dopamine agonists, MAO B inhibitors, catechol-O-methytransferase inhibitors, anticholinergics, amantadine, deep brain stimulation; stem cell introduction into brain.

Partington syndrome (XR) Intellectual disability, muscle spasm/tremors/uncontrolled movements of hands (dystonia), sometimes of face, speech, and walking. Seizures in some. **Rx**: Medication for dystonia; physical/speech therapy, anti-seizure medication if needed, special education.

Peters anomaly (AR, AD, sporadic) Bilateral cloudy corneas, low vision, glaucoma, cataract, microphthalmia. **Rx**: Corneal transplant, treatment of glaucoma, cataract surgery if needed.

Peutz-Jeghers syndrome (AD) Multiple gastrointestinal polyps. **Rx**: Removal of polyps.

Pfeiffer syndrome (AD) Premature fusion of skull bones, facial features, hearing loss, dental problems, brachydactyly (shortened fingers/toes), webbed digits. **Rx**: Reconstructive surgery; dental correction.

Phenylketonuria (AR) Intellectual disability, seizures, delayed development, musty odor from excess phenylalanine, skin disorders. **Rx**: Limit phenylalanine in diet.

Pheochromocytoma (AD) Adrenal medullary tumor, tachycardia, hypertension, cafe-au-lait spots. **Rx**: Surgery to remove tumor.

Phosphofructokinase deficiency (Type VII glycogen storage disease) (AR) Glycogen accumulation. Exercise intolerance, muscle pain/cramps. **Rx**: Avoid strenuous exercise; low-carbohydrate diet.

Phosphoglycerate kinase deficiency (AR) Anemia, enlarged spleen, intellectual dysfunction, epilepsy, weakness, muscle cramps, exercise intolerance, myoglobin in urine. **Rx**: Blood transfusion/splenectomy in some.

Phosphoribosylpyrophosphate (PRPP) superactivity (XR) Overproduction of uric acid; gout, kidney/bladder stones; some may have sensorineural hearing loss, ataxia, developmental delay. **Rx**: Uric acid-lowering agents, potassium citrate to alkalinize the urine, low-purine diet, high fluid intake.

PLAID (PLCG2-associated **A**ntibody deficiency and **I**mmune **D**ysregulation**) syndrome** (AD) Allergic skin reaction (urticaria) to cold, risk of infections; autoimmune thyroiditis, vitiligo (patchy loss of skin pigmentation) in some. **Rx**: Antihistamines, avoid cold triggers.

Plasma pseudocholinesterase deficiency (AR) Prolonged respiratory paralysis during general anesthesia, since succinylcholine cannot be broken down quickly. **Rx**: Respiratory support for hours until succinylcholine is cleared.

Polycystic kidney disease (AD, AR, or new mutation) Risk of kidney failure, high blood pressure, back pain, urinary tract infections, kidney stones, heart valve abnormalities, aortic aneurysm. **Rx**: Pain/blood pressure medication, antibiotics for urinary infections, low-sodium diet, diuretics, surgery to drain renal cysts.

Pompe disease (AR) Weakness, hepatomegaly, breathing/feeding problems, respiratory infections, hearing loss. **Rx**: Enzyme replacement (alglucosidase alfa) IV. Specialty care for individual symptoms.

Porphyria (AD, AR, XD) Abdominal pain, weakness, seizures, fever, breathing problems, anemia, splenomegaly, abnormal liver function. **Rx**: Intravenous heme or glucose infusions to reduce porphyrin production in liver, high-carbohydrate diet; splenectomy/liver and bone marrow transplant in some.

Potocki-Lupski (17p11.2) syndrome (AD, most cases arise de novo) Weak muscle tone, swallowing/feeding problems, failure to thrive, heart problems, delayed development, intellectual disability, behavioral problems, autism spectrum disorder, vision/hearing problems, facial features, other abnormalities. **Rx**: Supportive, multidisciplinary evaluation, cardiac consultation as needed.

Potocki-Shaffer syndrome (AD) Multiple benign bone tumors, persistent parietal foramina, facial features, vision problems, intellectual disability, autism spectrum disorder, other abnormalities. **Rx**: Surgery, ophthalmologist for visual problems, genetic counseling.

Prader-Willi syndrome (AD, loss of paternal allele) Insatiable appetite, obesity, delayed development. **Rx**: Good nutrition, human growth hormone, sex hormone treatment, weight management, treatment of sleep problems.

Preaxial polydactyly (AD) Extra finger/toe. **Rx**: Corrective surgery.

Progressive myoclonus epilepsy (AR, mitochondrial forms) Seizures (often triggered by light flashes, emotions), intellectual decline usually beginning in late childhood or adolescence. **Rx**: Supportive; anti-seizure medication not as effective as in other forms of epilepsy.

Propionic acidemia (AR) Poor feeding, hypotonia, lethargy, heart abnormalities, seizures; delayed development and intellectual disability in some; carnitine deficiency. **Rx**: Low-protein diet, carnitine supplementation, physical/occupational therapy, remedial education, hemodialysis if excess ammonia/acidosis.

Prostate cancer, hereditary (AD or somatic mutation). **Rx**: Surgery, radiation, cryotherapy, hormone therapy, chemotherapy.

Protoporphyrinogen oxidase deficiency (Variegate porphyria) (AD) Skin photosensitivity with blistering, neurogenic attacks of abdominal pain, nausea, vomiting, trouble urinating, discolored urine from porphyrins. **Rx**: Pain control, correct electrolyte imbalance; hemin administration (represses the heme pathway and helps recovery from the attack); avoid sunlight.

PRPP synthetase overactivity (XR) High uric acid in blood and urine. Gout, uric acid kidney stones. Some may have hearing loss, muscle weakness, ataxia, developmental delay. **Rx**: Uric acid-lowering drugs; low-fat, complex carbohydrate diet. Avoid high-purine foods.

Pseudohypoparathyroidism (AD) Short stature, lethargy, impaired intelligence, cataracts. **Rx**: Intravenous calcium, vitamin D metabolites.

Purine nucleoside phosphorylase deficiency (AR) Immune deficiency, infections, development delay, intellectual disability, ataxia, spasticity, risk of autoimmune disorders. **Rx**: Bone marrow transplant.

Pyruvate dehydrogenase deficiency (AR, X-inactivation) Lactic acidosis, nausea, vomiting, breathing difficulty, abnormal heartbeat, abnormal eye movements, seizures; carnitine deficiency. **Rx**: Thiamine, carnitine, lipoic acid help some. Low-carbohydrate, high-fat diet. Oral citrate for acidosis.

Pyruvate kinase deficiency (AR) Hemolytic anemia, jaundice, fatigue, shortness of breath, tachycardia, splenomegaly, gallstones. **Rx**: Blood transfusion, splenectomy in some to reduce red blood cell destruction.

Rapadilino syndrome (AR) Absent or underdeveloped bones of forearms/thumbs/kneecaps, cleft palate, feeding problems, short stature, cafe-au lait spots in some, risk for osteosarcoma/lymphoma. **Rx**: Orthopedic/nutritional management, watch for osteosarcoma.

Refsum disease (AR) Defect in lipid metabolism. Vision problems, poor muscle coordination, peripheral neuropathy, night blindness, dry rough skin. **Rx**: Diet low in phytanic acid and high in calories. Plasmapheresis in some.

Renal cysts and diabetes syndrome (AD) Abnormal renal development; maturity-onset diabetes of the young. **Rx**: Treatment of renal problem and diabetes.

Respiratory distress syndrome (Multifactorial) Labored breathing, weakness, low blood pressure, cough, fever, headaches, fast pulse, discolored skin/nails. **Rx**: Oxygen; fluid balance; medication for pain, infection, anticoagulants. Pulmonary rehab.

Retinitis pigmentosa (AD, AR, XR) Retinal degeneration, vision loss. **Rx**: Vitamin A palmitate may slow vision loss; low vision aids, cataract surgery for some; gene therapy, stem cell therapy, growth factors, retinal implants.

Rett syndrome (XD, usually no family history) Almost all in girls, problems with learning, coordination, repeated hand motions, staring, microcephaly, breathing problems, scoliosis, seizures, speech problems. **Rx**: Medication for seizures, digestive problems, heart rhythm abnormalities, breathing problems; physical/occupational/ speech therapy.

Reye syndrome (Associated with certain inborn errors of metabolism) Swelling of liver and brain, seizures, confusion, linked with aspirin use in flu or chickenpox. **Rx**: Steroids for brain swelling, diuretics for excess fluid; prevent bleeding with vitamin K, plasma and platelets, IV glucose and electrolytes.

Rhabdomyosarcoma (AR or somatic mutation) Striated muscle tissue cancer. **Rx**: Surgery, chemo/radiation therapy.

Roberts syndrome (AR) Limb and facial abnormalities, developmental delay, intellectual impairment. **Rx**: Surgical correction of facial/limb defects; limb prostheses for absent limbs; supportive treatment.

Rothmund-Thomson syndrome (AR) Rash, skin atrophy, telangiectases (dilated spidery capillaries on the skin or organ surface), sparse hair, small stature, gastrointestinal problems, cataracts, skeletal abnormalities, cancer risk (osteosarcoma) **Rx**: Sunscreen; keratolytics/retinoids for skin lesions; cataract surgery, screen for osteosarcoma.

Rotor syndrome (AR) Non-itching jaundice. **Rx**: No treatment needed.

Rubinstein-Taybi syndrome (AD) Short stature, intellectual disability, facial features, broad thumbs/large toes, risk of benign brain and skin tumors, abnormalities of eye, heart, kidney, teeth, obesity. **Rx**: Surgery to correct thumb bones; physical therapy, speech therapy, specialists for individual problems.

Sanfilippo A and B diseases (AR) Lysosomal storage disease. Buildup of mucopolysaccharides in lysosomes. Slow development, intellectual decline to dementia, temper tantrums, aggressive behavior, sleep disturbance, gradual immobility and unresponsiveness, swallowing difficulty, seizures; other neurologic, skeletal abnormalities. **Rx**: Supportive. Gene therapy/enzyme replacement trials.

Sarcopenia (Mitochondrial, multifactorial) Loss of skeletal muscle mass and function with aging. **Rx**: Exercise, strength training, testosterone/growth hormone supplements.

SBBYS (Say-Barber-Biesecker-Young-Simpson) syndrome (AD) Male genital anomalies, intellectual disability, facial features, underdeveloped kneecaps, other anomalies. **Rx**: Specialty treatment for individual problems.

Scheie disease (AR) Lysosomal storage disease. Skeletal abnormalities, delayed motor development, facial features, glaucoma, corneal opacification, hearing loss in some. **Rx**: Physiotherapy, enzyme replacement (laronidase), stem cell transplantation in some.

Schimke immuno-osseous dysplasia (AR) Short stature, immune deficiency, kidney disease, lordosis, hyperpigmented patches on chest and back. **Rx**: Kidney dialysis/transplant if needed; bone marrow transplant for immune deficiency; low white count may be treated with granulocyte colony-stimulating factor or granulocyte-macrophage colony-stimulating factor; immune suppression if disease manifests as autoimmune; hip replacement; scoliosis correction; antibiotics.

Schinzel-Giedion syndrome (AD) Severe congenital neurological, organ, and organ problems; midface retraction; developmental delay; tumors in some. **Rx**: Symptomatic, hearing aids, tumor treatment may include surgery/chemotherapy.

Schizophrenia (AD in some, but inheritance generally not known). **Rx**: Antipsychotic medications.

Shwachman-Diamond syndrome (AR) Bone marrow dysfunction, too few white cells/red cells/platelets resulting in infections/anemia/bleeding tendency. Risk of acute myeloid leukemia; digestion problems due to pancreatic insufficiency; skeletal abnormalities, short stature, delayed development, other organ system involvement. **Rx**: Pancreatic enyzme supplementation; treat infections, orthopedic, and hematologic problems.

Scleroderma (AD or multiple factors) Autoimmune fibrosis of skin, other organs, Raynaud phenomenon (white/blue fingers on cold exposure). **Rx**: Steroid creams or pills for skin problems; vasodilators to expand blood vessels in skin, lung and kidney; immune suppression; antacids; antibiotics; analgesics.

Senior-Loken syndrome (AR) Nephronophthisis (kidney cysts that impair kidney function), Leber congenital amaurosis (retinal problems, farsightedness, nystagmus). **Rx**: Monitor kidney function; dialysis/kidney transplant in some; visual aids; genetic counseling.

Septo-optic dysplasia (AR, AD) Brain anomalies – underdeveloped optic nerves/pituitary gland/brain midline, nystagmus, pituitary insufficiency, short stature, genital abnormalities. **Rx**: Symptomatic, hormone replacement,

Sertoli cell-only syndrome (Y-linked) Male infertility, low or absent sperm count. **Rx**: Sperm cell extraction from testes for fertilization may be helpful in some.

Short rib-polydactyly syndrome (AR) Extra fingers, small thorax with short ribs and cardiorespiratory problems, other bony defects, polycystic kidneys, transposition of large vessels, narrowing of GI and genitourinary system passages. Lethal. **Rx**: Supportive.

Sickle cell anemia (AR) Shortness of breath, fatigue, jaundice, painful sickling episodes, pulmonary hypertension. **Rx**: Avoid pain episodes; analgesics; special meds: hydroxyurea increases fetal hemoglobin, retards sickling, and reduces white blood cell sticking to blood vessels; L-glutamine to increase red blood cell flexibility; crizanlizumab to decrease red blood cell adherence; voxelotor to increase affinity of hemoglobin for oxygen and decrease sickling; blood transfusions, stem cell transplant in some, prevent infections.

Siderius XLMR (XR) Intellectual disability; delayed walking, speech, cleft palate; facial features. **Rx**: Genetic counseling, supportive.

Silver-Russell syndrome (AD, AR, mostly sporadic) Failure to thrive, short height, facial features, digestive system problems, curved fifth finger. **Rx**: Supportive, nutrition schedule, speech therapy, orthopedic correction.

Sjogren syndrome (Multifactorial) Autoimmune disease with dry eyes and mouth, sometimes with inflammation elsewhere. Rx: Artificial tears, occlusion of lacrimal punctum, anti-inflammatory agents.

Skin cancers (5-10% AD for melanoma skin cancer) **Rx**: Freezing, surgery, radiation, chemotherapy, photodynamic therapy depending on specific tumor.

Small intestine cancer (Increased risk with *familial adenomatous polyposis, hereditary nonpolyposis colorectal cancer, Peutz-Jeghers syndrome, cystic fibrosis, multiple endocrine neoplasia, Gardner syndrome, von Recklinghausen's disease*) **Rx**: Surgery, radiation, chemotherapy.

Smith-Lemli-Opitz syndrome (AR) Facial features; microcephaly; intellectual disability; autism; hypotonia; syndactyly; polydactyly; malformations of heart, lungs, kidneys, GI tract, genitalia; deficiency in cholesterol synthesis. **Rx**: Cholesterol supplementation.

Smith-Magenis syndrome (Spontaneous deletion of small part of chromosome 17) Facial features, intellectual disability, delayed speech, affectionate persona, temper tantrums, self-injury, self-hugging, short stature, scoliosis, daytime sleeping, nighttime awakenings, other abnormalities. **Rx**: Symptomatic, physical/occupational/speech therapy. Melatonin for sleep; medication (e.g. risperdal) for violent behavior.

Soft tissue sarcoma (Most not inherited, higher risk in family cancers, such as *Li-Fraumeni syndrome, neurofibromatosis, Gardner syndrome, Gorlin syndrome, tuberous sclerosis*). **Rx**: Surgery, radiation, chemotherapy.

Sotos syndrome (AD, 95% spontaneous) Facial features, delayed development, learning and motor disabilities, large head, tall, ADHD, speech problems. **Rx**: Behavioral/occupational/speech therapy, counseling, medication for ADHD, hearing aids/glasses as needed.

Spermatogenic failure, nonobstructive (Y-linked) Low or absent sperm count. **Rx**: Sperm extraction from testes for fertilization.

Spinal muscular atrophy (AR) Weakness, proximal muscle atrophy. **Rx**: Nusinersen and onasemnogene abeparvovec-xioi gene therapies to instruct genes to make a protein that helps control muscle movement; exercise; physical therapy.

Spinocerebellar ataxias (AD) Coordination/balance problems, speech/swallowing difficulty, spasticity, eye muscle weakness, nystagmus, cognitive impairment, sensory neuropathy, muscle atrophy. **Rx**: Symptomatic, rehab, n-acetyl-leucine to improve vestibular imbalance/nystagmus.

Stomach cancer (AD has disposition; about 10% familial) **Rx**: Surgery, radiation, chemotherapy, immunotherapy.

Tangier disease (AR) Reduced levels of high-density lipoprotein (HDL); hypertriglyceridemia; increased cardiovascular disease risk; neuropathy; large orange-colored tonsils. Some may have spleno/hepatomegaly, corneal clouding, Type II diabetes. **Rx**: Low-fat diet, avoid risk factors (smoking, high blood pressure, diabetes, obesity). Fibrates to lower triglycerides.

Tatton-Brown Rahman syndrome (AD) Facial features, tall, large head, intellectual disability, seizures, scoliosis. **Rx**: Physiotherapy, psychological support.

Tauopathies (AD, AR) Deposition of tau protein in nerve/glial cells, causing a variety of neurodegenerative disorders. **Rx**: Dopamine agonists for Parkinsonism; antidepressants; cholinesterase inhibitors; N-methyl-D-aspartate receptor antagonist for learning and memory; speech/physical therapy.

Tay-Sachs disease (AR) Starting 3-6 months, slow development, progressive weakness, vision/hearing loss, intellectual disability, paralysis, cherry-red spot on retina; death in early childhood. **Rx**: Palliative care.

Testicular cancer (Multiple genes—most cases have no family history). **Rx**: Surgery, radiation, chemotherapy, sometimes with pre-stored stem cell transplant to replace marrow cells damaged by chemotherapy.

Thalassemia (AR, AD) Anemia, paleness, weakness, fatigue, blood clots, jaundice; enlarged spleen, liver, heart in some; bone abnormalities. **Rx**: Blood transfusion, iron chelation therapy, folic acid supplements.

Thanatophoric dysplasia (AD) Short limbs, excess skin folds on arms and legs, enlarged head, wide-spaced eyes, underdeveloped lungs, usually stillborn or die shortly after. **Rx**: Palliative care.

Thyroid cancer (AD, somatic mutation) **Rx**: Surgery, radioactive iodine, thyroid hormone replacement, radiotherapy, chemotherapy, targeted therapy.

Transient neonatal diabetes mellitus (AR, random mutations) **Rx**: Insulin, glibenclamide (stimulates pancreatic insulin release).

Trichothiodystrophy (AR) Brittle and sparse hair, dry scaly skin, short stature, slow growth, common intellectual disability, recurrent respiratory infections, congenital cataracts, poor coordination, skeletal anomalies. **Rx**: Manage symptoms, sun protection if photosensitive.

Trifunctional protein deficiency (AR) Mitochondrial difficulty converting fats to energy, fatigue, feeding difficulty, hypoglycemia, hypotonia, liver/heart/pulmonary problems; carnitine deficiency. **Rx**: Low-fat diet; carnitine; substitute medium chain fatty acids for long chain fatty acids; avoid fasting, environmental extremes, and excess exercise.

Triosephosphate isomerase deficiency (AR) Anemia, movement disorder, infections, weakness, breathing/heart problems. **Rx**: Blood transfusions, assisted ventilation, genetic counseling, otherwise supportive therapy.

Turner syndrome (Most not inherited, monosomy X) Short stature, ovarian hypofunction, webbed neck, lymphedema of hands and feet, skeletal/kidney problems, coarctation of aorta. **Rx**: Growth hormone; estrogen replacement; attention to ear infections, high blood pressure, and thyroid disorders.

Tyrosinemia I (AR) Can't break down tyrosine. Failure to thrive, jaundice, cabbage-like odor, intellectual disability, bleeding tendency, liver/kidney failure, rickets, liver cancer, some with peripheral neuropathy, respiratory problems. **Rx**: Dietary restriction of tyrosine and phenylalanine; nitisinone to help prevent high blood levels of tyrosine.

Uroporphyrinogen decarboxylase deficiency (porphyria cutanea tarda) (AD, AR) Photosensitivity with skin blisters, hyperpigmentation, dark urine, excess liver iron, liver damage, risk of cirrhosis/liver cancer. **Rx**: Avoid sunlight, alcohol; phlebotomy/chelators to remove iron from blood.

Uroporphyrinogen I deficiency (acute intermittent porphyria) (AR) severe abdominal pain; pain also in chest, legs, back, nausea/vomiting, muscle pain/tingling/weakness; dark urine; high blood pressure; seizures; mental changes (hallucinations, paranoia, anxiety, confusion). **Rx**: Hospitalization may be necessary to treat pain, kidney failure, and/or liver damage.

Uroporphyrinogen III cosynthase deficiency (congenital erythropoetic porphyria) (AR) Reddish discoloration of urine and teeth; light sensitivity with skin blisters and scarring; excess hair growth. **Rx**: Blood transfusion, bone marrow/hematopoietic stem cell cord transplantation, blood transfusion, splenectomy in some to reduce amount of heme production.

Usher syndrome (AR) Progressive vision/sensorineural hearing loss. Retinitis pigmentosa; cataracts; balance problems in some. **Rx**: Manage vision, hearing, balance problems. Hearing aids; cochlear implants; low-vision devices; mobility training.

Uterine cancer (Increased risk with AD BRCA1, BRCA2 mutations or Lynch syndrome) **Rx**: Surgery, radiation, chemotherapy, hormone therapy.

Van Buchem disease (AR) Dense bones, enlarged jaw, misaligned teeth, facial features, headaches, cranial nerve compression (facial paralysis, hearing/vision loss, decreased smell, brainstem compression), may have webbed fingers. **Rx**: Surgical correction of mandible, cranial nerve decompression.

Very-long-chain acyl-CoA dehydrogenase deficiency (AR) Disorder of fatty acid metabolism. Lethargy in infancy, cardiomyopathy, hypoglycemia. **Rx**: Low-fat, high-carbohydrate diet, with supplementary medium chain triglycerides; symptomatic treatment; genetic counseling.

Von Gierke disease (GSD Type I) (AR) Glycogen accumulation in liver, kidneys, small intestines impairs their function. Hypoglycemia, seizures, enlarged liver, xanthomas (cholesterol deposits in skin), osteoporosis, gout, high blood pressure, pulmonary hypertension, polycystic ovaries, liver tumors, low white cell count with infection susceptibility. **Rx**: Diet to control hypoglycemia; uric acid/lipid-lowering medication; granulocyte colony stimulating factor for recurrent infections; surgery for liver tumors; kidney/liver transplantation where indicated.

Von Hippel-Lindau syndrome (AD, but need second mutation to cause tumor/cyst formation) Tumors and cysts in various body areas. Tumors can be benign (e.g. hemangioblastoma, pheochromocytoma) or malignant (e.g. clear cell carcinoma). Cysts in kidneys, pancreas, genital tract. **Rx**: Surgery, radiation.

Waardenburg syndrome (AD, AR, or spontaneous mutations) Sensorineural hearing loss; pigmentation changes in hair (e.g. white patches), skin, and eyes (e.g. may have one blue and one brown iris). **Rx**: Hearing aids.

Warsaw breakage syndrome (AR) Impaired growth, short stature, microcephaly, intellectual disability, facial features, hearing loss, heart malformations. **Rx**: Supportive.

Werner syndrome (AR) Premature aging starting in teens and twenties, bird-like face, cataracts, arteriosclerosis, osteoporosis, predisposition to some types of cancer. **Rx**: Surgery, chemotherapy, radiation, genetic counseling.

Williams-Beuren syndrome (AD) Developmental delay, intellectual disability, facial features, heart/cardiovascular problems, hypertension, hypercalcemia, hernia, narrowing of large blood vessels, friendly talkative persona, good memory for songs and sounds. **Rx**: Symptomatic; physical/speech therapy; surgery for narrowed blood vessels in some.

Wilms tumor (AD, XD, but 90% are somatic mutation) Most common kidney cancer in children. **Rx**: Surgery, chemotherapy, radiation.

Wilson disease (AR) Often noticed first in teenage years. Excess copper accumulation, especially in liver, brain, eyes (Kayser-Fleischer copper ring around cornea). Jaundice, fatigue, ataxia, speech difficulty, cognitive disability, mood swings. **Rx**: Chelating agents to remove excess copper; liver transplant in some.

Wiskott-Aldrich syndrome (X-linked) Immune deficiency, eczema, decreased platelets with bleeding tendency, decreased white cells with increased susceptibility to infections, autoimmune disorders (e.g. rheumatoid arthritis, vasculitis, hemolytic anemia), lymphoma. **Rx**: Antibiotics, IV immunoglobulin, splenectomy in some, gene therapy, stem cell transplantation, medication for immune system regulation.

Wolf-Hirschhorn syndrome (85-90% non-inherited; translocation) failure to thrive, delayed development, intellectual disability, facial abnormalities, seizures, skeletal/dental problems. **Rx**: Physical/occupational therapy, surgery for physical defects.

Xanthine oxidase deficiency (AR) Excretion of urinary xanthine and hypoxanthine. May be asymptomatic or have renal dysfunction with xanthine kidney stones. **Rx**: High fluid intake, low-purine diet, alkalinization of urine.

X-linked autoimmunity-allergic dysregulation syndrome (XR) Autoimmunity (Type I diabetes, under- or overactive thyroid, anemia, low platelets), polyendocrinopathy, intestinal dysfunction, dermatitis. **Rx**: Medication to modulate immune system, bone marrow transplant.

X-linked deafness (X) **Rx**: Supportive, hearing aids, cochlear implant.

X-linked dyserythropoietic anemia and thrombocytopenia (X) Anemia, thrombocytopenia, bleeding, susceptibility to infections. **Rx**: Platelet transfusions, desmopressin in some (stimulates release of von Willebrand clotting factor), bone marrow transplantation in some.

X-linked lissencephaly with abnormal genitalia (XR) Mostly affects males. Smooth brain folds, absent corpus callosum in some, micropenis, ambiguous genitalia, developmental delay, intellectual disability, spasticity, seizures. **Rx**: Symptomatic, supportive, anticonvulsant drugs.

X-linked severe combined immunodeficiency ("bubble boy" disease) (XR) Almost all in males. Recurrent infections, diarrhea, thrush. **Rx**: Plastic enclosure to protect against infection; antibiotics, intravenous immunoglubulin supplementation, bone marrow transplant, gene therapy to replace defective gene (risk of leukemia side effect).

X-linked thrombocytopenia (XR) Mainly affects males. Shortage of platelets; bleeding, eczema; prone to infections; cancer or autoimmune problems in some. **Rx**: Antibiotics, platelet transfusions; possibly stem cell transplant, corticosteroids, IV immunoglobulin.

Xeroderma pigmentosum (AR) Sensitivity to sunlight, susceptibility to skin (and other) cancers. May have neurologic problems (ataxia, loss of intellectual function, difficulty hearing/swallowing/talking, seizures). **Rx**: Sunscreen; surgery for skin cancers; isotretinoin or acitretin to relieve keratoses.

Index